Thorsten U. Kampen

Low Molecular Weight Organic Semiconductors

Related Titles

Yersin, H. (ed.)

Highly Efficient OLEDs with Phosphorescent Materials

458 pages with approx.
195 figures and approx. 40 tables
2008
Hardcover
ISBN: 978-3-527-40594-7

Hadziioannou, G.,
Malliaras, G. G. (eds.)

Semiconducting Polymers
Chemistry, Physics and Engineering

768 pages in 2 volumes
with 402 figures and 3 tables
2007
Hardcover
ISBN: 978-3-527-31271-9

Klauk, H. (ed.)

Organic Electronics
Materials, Manufacturing and Applications

446 pages with 281 figures and 21 tables
2006
Hardcover
ISBN: 978-3-527-31264-1

Müllen, K., Scherf, U. (eds.)

Organic Light Emitting Devices
Synthesis, Properties and Applications

426 pages with 258 figures and 14 tables
2006
Hardcover
ISBN: 978-3-527-31218-4

Brütting, W. (ed.)

Physics of Organic Semiconductors

554 pages with 323 figures
2005
Hardcover
ISBN: 978-3-527-40550-3

Thorsten U. Kampen

Low Molecular Weight Organic Semiconductors

WILEY-VCH Verlag GmbH & Co. KGaA

The Author

PD Dr.rer.nat. Thorsten U. Kampen
SPECS GmbH, Berlin, Germany
TU Berlin, Germany

Library of Congress Card No.:
applied for

British Library Cataloguing-in-Publication Data
A catalogue record for this book is available from the British Library.

Bibliographic information published by the Deutsche Nationalbibliothek
The Deutsche Nationalbibliothek lists this publication in the Deutsche Nationalbibliografie; detailed bibliographic data are available on the Internet at http://dnb.d-nb.de.

Cover Design Adam-Design, Weinheim
Typesetting Toppan Best-set Premedia Limited
Printing and Binding T.J. International Ltd., Padstow, Cornwall

Printed in Great Britain
Printed on acid-free paper

ISBN: 978-3-527-40653-1

Contents

Low Molecular Weight Organic Semiconductors. Thorsten U. Kampen

ISBN: 978-3-527-40653-1

Preface

This book has covers the physical properties of organic and their application in organic/inorganic interfaces. The investigations have been performed using complementary experimental and theoretical techniques. Here, the collaborations within the European Research Training Network known as DIODE (Designing Inorganic Organic Devices, HPRN-CT-1999-00164) have played an important role.

In this respect, I have to thank Iggy McGovern, Justin Wells, and Gregory Cabailh from the Trinity College in Dublin, Ireland, as well as Andrew Evans, Alex Veary-Roberts, and Adam Bushell from the University of Wales in Aberystwyth for their cooperation during synchrotron beamtimes at BESSY I and II in Berlin. Javier Mendez from Universidad Autonoma de Madrid, Spain, performed the atomic force microscopy investigations shown in this work. Concerning the theoretical part of this work, I have to thank Fernando Flores from the Universidad Autonoma de Madrid for a long and fruitful cooperation in the field of semiconductor surfaces and interfaces. The simulations of the current-voltage characteristics of organic modified Schottky contacts have been done in collaboration with Aldo DiCarlo from Universita degli Studi di Roma "Tor Vergata" in Italy.

I want to thank Walter Braun, Torsten Kachel, Patrick Bressler, and David Batchelor from BESSY in Berlin, Germany, for their support during our synchrotron beamtimes. The research at this facility was funded by the Bundesministerium für Bildung und Forschung in Germany (BMBF contract No. 05 KS1OCA/1).

Thanks are due to Norbert Karl (University of Stuttgart, Germany) and Antoine Kahn (Princeton University, USA) for long and intense discussions on the properties of organic semiconductors and their interfaces.

I had a wonderful and exciting time during my stay at the University of Nagoya, Japan, in 2000. Kazuhiko Seki and Hisao Ishii took care of me in every aspect, and since then, the research groups in Nagoya and Chemnitz have a fruitful collaboration in vibrational spectroscopy investigations on organic/metal interfaces.

My special thanks go to the former and momentary members of the Semiconductor Physics group at Chemnitz University of Technology. I am indebted to: Arindam Das, Kornelia Dostmann, Axel Fechner, Marion Friedrich, Gianina Gavrila, Pham Truong Giang, Mihaela Gorgoi, Cameliu Himcinschi, Stefan Hohenecker, Andrei Kobitski, Martin Lübbe, Thomas Lindner, Henry Mendez,

Low Molecular Weight Organic Semiconductors. Thorsten U. Kampen

ISBN: 978-3-527-40653-1

Andreea Paraian, Sunggook Park, Sybille Raschke, Georgeta Salvan, Dmitri Tenne, and Ilja Thurzo.

Wolfram Fliegel from Freiberger Compound Materials GmbH in Freiberg, Germany, supplied GaAs wafers.

Since I started working with organic materials, Reinhard Scholz (TU Muenchen Germany) has been always a willing confidant when discussing all aspects of the physical properties of organic materials. I am especially indebted to Dietrich R. T. Zahn for giving me the opportunity to work in his group.

That I was able to complete this work at all is a tribute to my beloved wife Ellen, who inspired and goaded me into keeping up the work. I have to thank her for her patience and never ending support.

My scientific approach to organic semiconductors is based on my scientific background before I started working in this field, that is, semiconductor surfaces and interfaces. This and the outline of research in the above mentioned DIODE network determines the outline of this book. In the DIODE network we focused on the modification of metal semiconductor interfaces using organic molecules. We had the unique opportunity to start from basic research on finding the proper prepared semiconductor surface for organic growth, then investigating the bulk and interface properties of the organic films grown on inorganic semiconductor substrates, and finally using the findings to optimise and test a real device.

Therefore, the book starts with an introduction into the surface and interface properties of inorganic semiconductor surface in Chapter 1. Chapter 2 and 3 present the techniques for the purification/growth of organic semiconductors and the resulting structural properties, respectively. The optical properties of such thin films are presented in Chapter 4, while Chapter 5 discusses the chemical and electronic surface properties. Finally, Chapter 6 presents charge transport properties. The experimental results presented are obtained with a variety of complementary techniques which are briefly introduced.

Thorsten U. Kampen

1
Introduction

The reports of efficient electroluminescence from conjugated polymers [1] and small molecules [2] have triggered spectacular developments in the field of organic-based electro-optic and electronic devices, such as optical switches [3], batteries [4], field-effect transistors and circuits [5, 6], identification and product tagging [5, 7], data storage and computing [8], sensors and actuators [9], solar cells [10], organic light-emitting diodes (OLEDs) [1], flat panel displays [11, 12], photodiodes, solar cells, solid-state lasers, integrated circuits, and waveguides [13–15]. The low synthesis costs and relative easiness of handling make this new class of materials attractive for the above-mentioned applications. Furthermore, the chemical compatibility of organic materials with plastics allows the low-cost fabrication of flexible, unbreakable, and transparent devices.

However, the low mobility of organic semiconductors precludes the complete replacement of compound inorganic materials, especially for high-frequency applications. On the other hand, organic semiconductor can be used in hybrid organic/inorganic devices to tailor properties and performances of conventional high-frequency devices. One of the possible applications of ultrathin layers of organic molecules is the passivation of semiconductor surfaces, such as Si(111)-(7 × 7) and GaAs(100) [16]. For example, a C_{60} monolayer on a Si(111)-(7 × 7) surface inhibits chemical contamination by water and atmospheric oxygen.

Another application is the organic modification of metal–semiconductor or Schottky contacts. Schottky diodes are often used for mixing applications in telecommunication systems, radio astronomy, radar technology, and plasma diagnostics [17]. One disadvantage is the relatively high bias voltage necessary to operate mixer diodes. This leads to high power consumption or low sensitivities combined with high requirements concerning the stability and noise of the voltage sources.

For all the applications mentioned earlier, thin film and interface properties are some of the most important issues with regard to the overall device concept and performance. The interface can influence the structural, optical, and electronic properties of organic films. The degree of order or disorder at the interface determines the growth mechanism, film morphology, and defect density in the film. Here, the chemical bonding between substrate and the first molecular layer will determine the molecular orientation in the first layer, and realignment of the molecules as the function of the film thickness may occur. The chemical bonding

Low Molecular Weight Organic Semiconductors. Thorsten U. Kampen

ISBN: 978-3-527-40653-1

will also influence the electronic interface structure, for example, via the formation of interface dipoles and/or band offsets. Highly ordered films may have improved transport properties due to the formation of a band structure, thus enabling coherent transport with large carrier mobilities and reduced scattering at defects or grain boundaries [18]. On the other hand, reduced defect and grain boundary densities in highly ordered films would reduce the number of paths for nonradiative recombination of electron–hole pairs, thus enhancing photoluminescence output.

It is tempting to use concepts for the description of organic semiconductor bulk and interface properties, which work well for inorganic semiconductor, but the experimental data available up to now already show that these concepts do not necessarily hold for organic semiconductors.

There are a large variety of molecules that fall in this class of organic semiconductors. They may be separated in two groups: polymers and so-called low molecular weight molecules. While polymers are long molecular chains, the group of low molecular weight molecules consists of molecules with about 10–100 atoms having a filled π-electron system. The experimental results presented here are obtained from three different kinds of organic molecules.

The perylene derivatives are commercial pigments and are derived from PTCDA. First perylene derivatives have been produced in 1912, but their application as pigments has started in 1950 after intensive research by Harmon Colors.

The synthesis of PTCDA is shown in Figure 1.1 [290]. It starts with the oxidation of acenaphthene using vanadium pentoxide as a catalyst. The resulting naphthalene dicarboanhydride reacts with ammonia to 1,8-naphtalene dicarboimide. This product converts in KOH at temperatures around 190–220 °C and subsequent oxidation of the melt in air into *N,N′*-3,4,9,10-perylenetetracarboximide (PTCDI). PTCDI is hydrolyzed in H_2SO_4 at 220 °C into PTCDA. Based on PTCDA, further perylene pigments are synthesized by exchanging the oxygen in the anhydride group with NR groups, where R is H, CH_3, or substituted phenylene. This is achieved by heating PTCDA at 150–200 °C in solvents containing aliphatic or aromatic amine. Using sulfuric or phosphoric acid can accelerate this reaction [19]. It should be mentioned that it is also possible to produce asymmetric substituted perylene pigments [20, 21].

Perylene pigments are used in high-quality varnishes in the car industry. Here, PTCDA and DiMe-PTCDI are known as Pigment Red 224 and Pigment Red 179, respectively. Their chemical structure and HOMO distribution are shown in

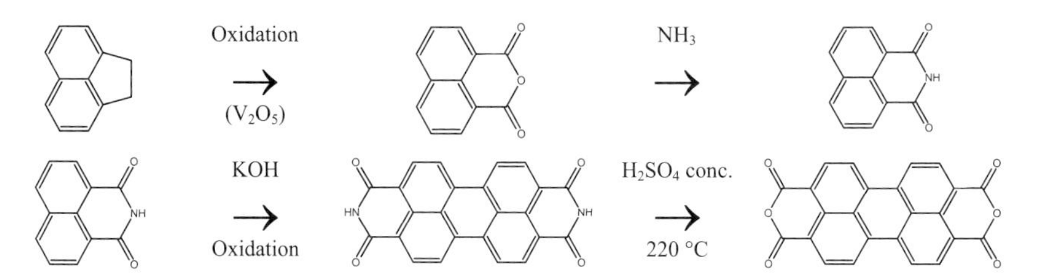

Figure 1.1 Synthesis of PTCDA.

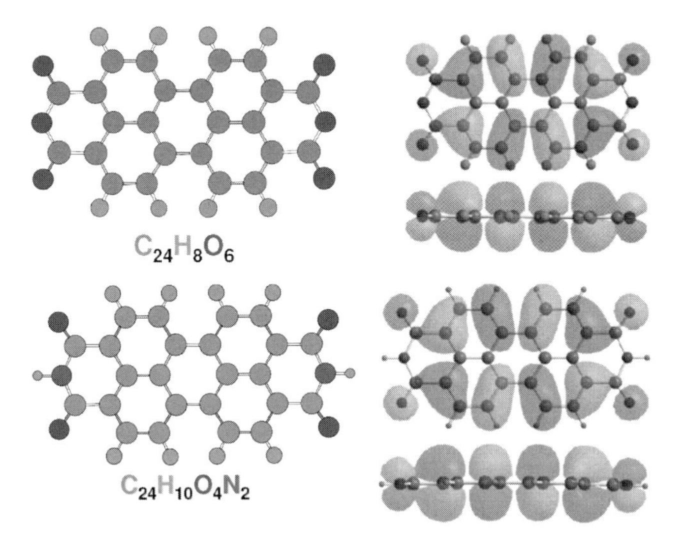

Figure 1.2 Molecular structure and HOMO distribution of PTCDA (top) and DiMe-PTCDI (bottom).

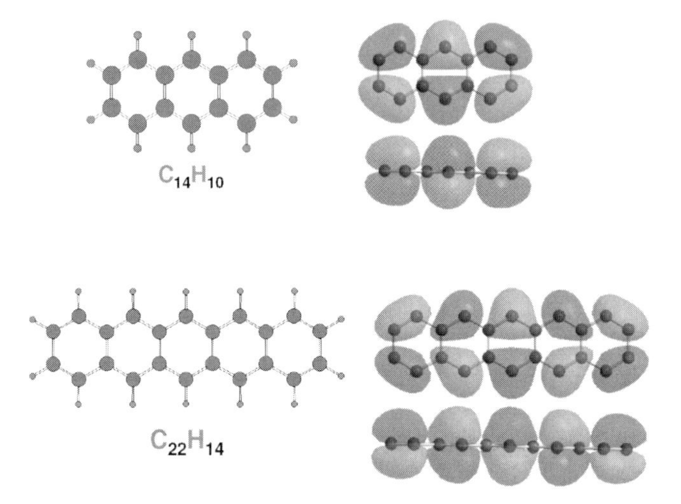

Figure 1.3 Molecular structure and HOMO distribution of anthracene (top) and pentacene (bottom).

Figure 1.2. They are thermally and photochemically stable. PTCDA dissociates at temperatures above 500 °C by losing its carboxylic groups. The molecular mass of PTCDA and DiMe-PTCDI amounts to 392 and 418 amu, respectively.

Linear acenes are polycyclic hydrocarbons consisting of linearly fused benzene rings. According to the number of benzene rings, the molecules are called anthracene (3), tetracene (4), and pentacene (5). Their respective structures are shown in Figure 1.3. Anthracene, its electrical conduction properties, and its application in

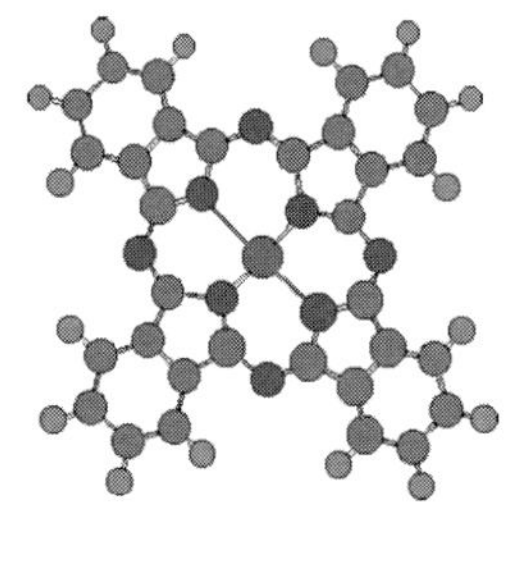
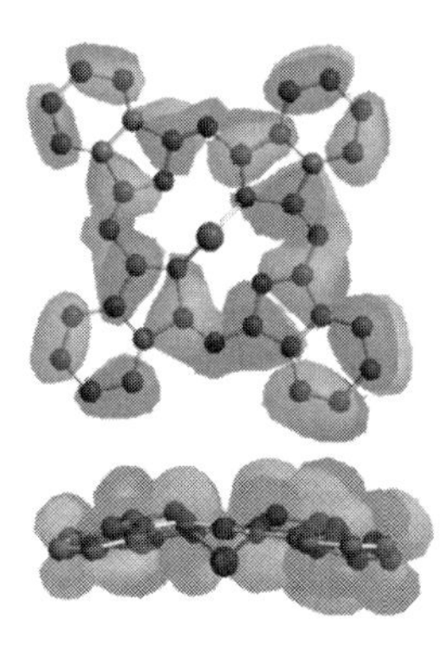

Figure 1.4 Molecular structure and HOMO distribution of copper phthalocyanine (CuPc).

photovoltaic cells have been studied about 100 years ago [22–24]. The molecules form perfect crystals, which are ideal samples for the investigations on charge carrier mobilities in organic materials. Pentacene, on the other hand, consists of five linearly fused benzene rings. Like anthracene, it shows high carrier mobilities.

Phthalocyanines have a principle fourfold symmetry and can coordinate different atoms in the center. As an example, Figure 1.4 shows the molecular structure of copper phthalocyanine (CuPc). The simplest case would be a hydrogen atom in the center, in which case the molecule is flat. Larger atoms like Sn will protrude on one side and the molecule will assume an umbrella-like shape. Phthalocyanines have been first synthesized at the beginning of the 20th century. Nowadays they are used in commercial dyes for the textile and paper industry. For this application, the solubility of the molecules is enhanced by attaching sulfonic acid functions. For example, the sodium salt of CuPc-sulfonic acid is known as Direct Blue 86.

1.1 Electronic Surface Properties of Inorganic Semiconductors

Adatoms on semiconductor surfaces generally induce changes in surface band bending as well as of the ionization energy. These effects are caused by adsorbate-induced surface states and surface dipoles, respectively. Surface band bending involves charge transfer from surface states into an extended space-charge layer beneath the surface, while adatom-induced surface dipoles are due to the partial ionic character of the covalent bonds between adatoms and surface atoms on the substrate. The latter effect may be described as a polarization of the bond charge toward the more electronegative atom. The direction of such bond-charge shifts determines the orientation of adatom-induced surface dipoles.

Already concentrations of surface states well below 1% of a monolayer will lead to what is called a *pinning of the Fermi level* on semiconductor surfaces. Such a Fermi level pinning and its associated band bending may persist in devices pre-

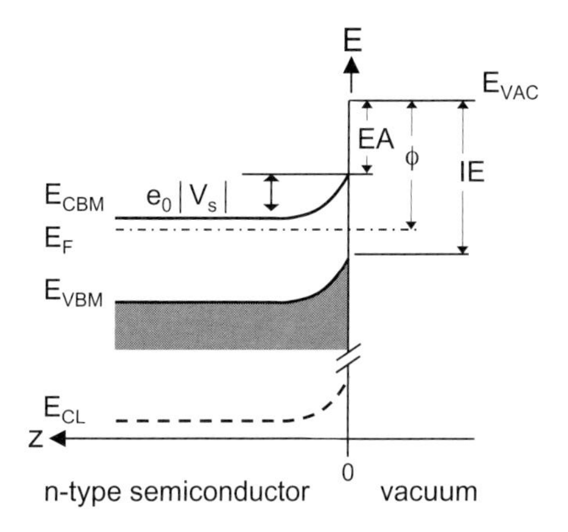

Figure 1.5 Electronic surface properties of an n-type doped semiconductor.

pared on such semiconductor surfaces resulting in injection barriers for charge carriers at semiconductor interfaces. To avoid these problems, special surface treatments are used to achieve what is called a surface passivation.

A detailed description of the properties of semiconductor surfaces and interfaces can be found in the monographs written by Lüth [25] and Mönch [26].

1.1.1 Surface Band Bending

Figure 1.5 illustrates the electronic surface properties of an n-type doped semiconductor. The most common surface property is the workfunction ϕ, which is defined as the difference between the Fermi energy E_F and the vacuum level E_{vac}:

$$\varphi = E_{vac} - E_F \tag{1.1}$$

In nondegenerated semiconductors, the Fermi level is positioned within the band gap, which is the energy region between the conduction band minimum E_{cbm} and the valence band maximum E_{vbm}. The energy gained by adding an electron is the electron affinity

$$EA = E_{vac} - E_{cbm} \tag{1.2}$$

while the ionization energy

$$IE = E_{vac} - E_{vbm} \tag{1.3}$$

is the minimum energy necessary to remove an electron from the solid. With Eq. (1.3), the workfunction of a semiconductor may be written as

$$\phi = IE - (E_F - E_{vbm}) \tag{1.4}$$

In Figure 1.5, an upward bending of the bands at the surface is assumed, which results in a different energy position of the valence band maximum $E_F - E_{vbm}$ (or conduction band minimum $E_F - E_{cbm}$) in the bulk and at the surface. The case shown here is a depletion layer, which is the most common situation for semiconductor surfaces. The result of this band bending $e_0 V_s$ is a space charge Q_{sc} carried by the positively charged ionized donor atoms, which are compensated by electrons in the bulk. According to the charge neutrality condition, the space charge compensates a surface charge Q_{ss} of equal magnitude but of opposite sign.

In general, screening of electrical charges in solids depends on the carrier density. The Debye length determines screening for semiconductors

$$L_D = \left[\varepsilon_b \varepsilon_0 k_B T \Big/ e_0^2 (n+p) \right]^{1/2} \tag{1.5}$$

with ε_b, n, and p are the dielectric constant of the semiconductor, the bulk electron, and hole concentration, respectively. For GaAs (ε_b = 12.85) [27] having a doping concentration of $1 \times 10^{15}\,\mathrm{cm}^{-3}$ ($1 \times 10^{17}\,\mathrm{cm}^{-3}$), the extrinsic Debye length amounts to 135 nm (13.5 nm). In metals, carrier screening is more efficient due to carrier densities being larger by several orders of magnitude than in nondegenerately doped semiconductors. Here, screening is determined by the Thomas–Fermi screening length

$$L_{\mathrm{Th-F}} = \left[\left(\frac{e_0^2}{\varepsilon_0} \right) D_m(E_F) \right]^{-1/2} \tag{1.6}$$

with $D_m(E_F)$ being the number of states per unit volume and unit energy at the Fermi level. For $D_m(E_F) = 3.45 \times 10^{22}\,\mathrm{cm}^{-3}\,\mathrm{eV}^{-1}$, which is the experimental value for aluminum, one obtains $L_{\mathrm{Th-F}} = 4 \times 10^{-2}$ nm, which amounts to only 14% of the nearest-neighbor distance of 0.286 nm in aluminum.

With the band bending $e_0 V_s$, the relation for the workfunction in Eq. (1.4) can be written as

$$\phi = IE + e_0 V_s - E_g + (E_{\mathrm{cbm}} - E_F)_b \tag{1.7}$$

Here, the subscript b indicates that the corresponding quantity has to be taken in bulk. Changes in the workfunction at constant temperature are due to changes in the ionization energy and/or the band bending:

$$\Delta\phi = \Delta IE + e_0 \Delta V_s \tag{1.8}$$

The surface charge at semiconductor surfaces Q_{ss} is due to intrinsic or extrinsic surface states. Intrinsic surface states are solutions of the Schrödinger's equation with complex wavevectors.

Figure 1.6 displays the complex band structure of a linear lattice. In the bulk complex, wavevectors are physically of no importance since Bloch waves would exponentially grow with $z \to \infty$ and cannot be normalized. For real k values, the well-known band structure is obtained with nearly parabolic dispersion and an energy gap at the boundary of the first Brillouin zone. In this sense, states with complex wavevector are called virtual gap states (ViGSs) of the complex band structure [28]. At surfaces, complex wavevectors become relevant. Here, they result in wave functions, which exponentially decay from the surface into the semicon-

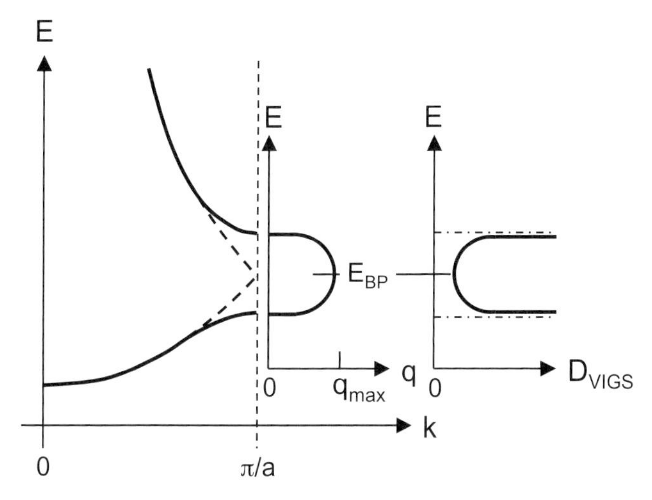

Figure 1.6 Complex band structure of a linear lattice and density of virtual induced gap states.

ductor and have an exponential tail decaying into the vacuum. The complex band structure contains an energy loop at the boundary of the Brillouin zone with complex wavevectors in the gap between the two bulk bands. The boundary condition for real surface states requires that their wave functions, which decay exponentially into the lattice, can be matched with an exponential tail into vacuum [29].

Surface states derive from the states in the bulk. Correspondingly, the character of surface states changes across the band gap from predominantly donor- to predominantly acceptor-like closer to the valence band maximum (VBM) and the conduction band minimum (CBM), respectively. The energy, at which the contributions from both bands are equal in magnitude, is called the branch point E_{bp}. In the band structure, the branch point is at the position where the complex wavevector has its maximum and the density of states has its minimum.

At clean semiconductor surfaces, the surface atoms have less than the four nearest- neighbor atoms in the bulk. On surfaces of elemental semiconductors like Si and Ge, which are terminated by a bulk lattice plane, each dangling bond should ideally contain one electron. This leaves nonsaturated or dangling bonds at the surface, which contain one electron and are responsible for surface states. In the case of zincblende-structure compound semiconductors, dangling bonds of cations and anions contain 3/4 and 5/4 of an electron charge, respectively. Occupied or nonoccupied surface states of acceptor character are negatively charged or neutral, respectively. Therefore, only acceptor surface states below the Fermi level are charged negatively, while donor surface states are positively charged above the Fermi level. The charge in acceptor surface states is compensated by a space charge of positive sign. This is achieved by an upward bending of the bands at the surface resulting in a region being depleted of electrons, as can be seen in Figure 1.7. The space charge is carried by the now uncompensated positively charged bulk donors. For surface states of donor-type character, the situation is

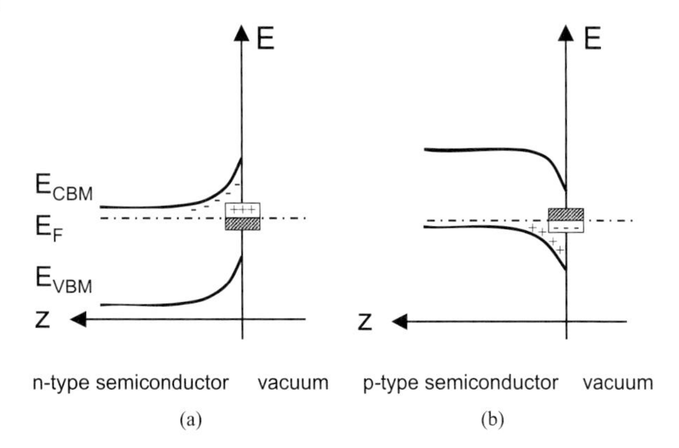

Figure 1.7 Charging of (a) acceptor- and (b) donor-type surface states on n and p-type semiconductor surfaces, respectively.

vice versa. Here, donor-type surface states result in band bending on p-type semiconductor surfaces.

The band bending results not only in the space charge region but also shifts the Fermi level closer to the respective surface states. The Fermi level comes close to the surface states for already very low surface states densities in the order of $10^{12}\,cm^{-2}$ and the surface band bending seems to be saturated. This behavior is often called *pinning of the Fermi level*. Changes in the band bending by a few k_BT cause only slight variations in the space charge, while occupancy of the surface states described by the Fermi–Dirac distribution function changes from one to zero. One has to keep in mind that the density of surface states sufficient for a pinning of the Fermi level is by orders of magnitude smaller than the number of adsorption sites on a semiconductor surface, which amounts to approximately $10^{15}\,cm^{-2}$.

Extrinsic surface states are formed by the adsorption of adatoms on semiconductor surfaces. For low surface densities, the adatoms will interact with their next-neighbor surface atoms via their dangling bonds. This will result in new surface states. Nonmetallic adatoms have been found to induce surface states of acceptor type on GaAs, where the band bending seems to correlate with the electronaffinity of the adatom [30]. Metal adatoms induce surface states of donor type [31, 32] and band bending scales with the atomic ionization energy of the adatoms [33].

1.1.2
Surface Dipoles

The charge transfer in covalent adsorbate substrate bonds on semiconductor surfaces leads to surface dipoles, which change the ionization energy. According to Pauling's concept [34], the ionic character Δq_1 of covalent single bonds in diatomic molecules may be described by the difference $X_A - X_B$ of the electronegativities of the two atoms involved. A revised version of Pauling's original correlation is [35]

$$\Delta q_1 = 0.16|X_A - X_B| + 0.035|X_A - X_B|^2 \tag{1.9}$$

The dipole moment of such molecules may then be written as

$$\mu_0 = \Delta q_1 e_0 \sum r_{\text{cov}} \tag{1.10}$$

where e_0 is the electronic charge and Σr_{cov} is the sum of the covalent radii of the atoms involved. Considering nearest-neighbor interaction between adatoms and surface atoms of the substrate only, Pauling's concept is easily applied to adatoms on semiconductor surfaces.

Adatom-induced surface dipoles may be described as an electric double layer. The voltage drop across this layer causes a change in ionization energy. The maximal variation of the ionization energy may be estimated for the maximal normal component of the dipole moment or, in other words, assuming the adsorbate substrate bonds to be perpendicular to the surface. Considering the mutual interaction between adatom-induced surface dipoles, the change in ionization energy is given by [36]

$$\Delta IE = -\frac{e_0}{\varepsilon_0} \frac{\mu_0 N_{\text{ad}}}{1 + 9\alpha_{\text{ad}} N_{\text{ad}}^{3/2}} \tag{1.11}$$

with ε_0 and α_{ad} being the permittivity of vacuum and the polarization of adatoms, respectively. Due to adatoms being electronegative or electropositive compared to the substrate surface atoms the ionization energy will increase or decrease, respectively.

1.1.3 Passivation of GaAs Surfaces

The passivation of semiconductor surfaces has two goals, namely the chemical and the electronic passivation. Chemical passivation means that the surface is inert against the absorption of foreign atoms or molecules. For example, in silicon device technology oxide layers are often removed by etching in hydrofluoric acid (HF) solutions. By using high-resolution energy loss spectroscopy [37] and infrared absorption spectroscopy [38], HF-treated surfaces were found to be hydrogen terminated. Such hydrophobic silicon surfaces are strongly passivated against interaction of the oxygen. The oxygen uptake takes place at exposures about 10–13 orders of magnitude larger than on clean Si(111)-(2 × 1) or Si(111)-(7 × 7) surfaces [39].

An electronic passivation should result in a flat band condition at the surface or, in other words, the Fermi level has the same energy position in the band gap of the semiconductor at the surface and in the bulk. This is achieved by the removal of all surface states within the band gap. Electronic passivation may be used as a method to control Schottky barrier heights [40]. It has been shown in different theoretical [41] and experimental works [42–45] that passivation tends to reduce barrier heights on n-type doped semiconductors, offering ways of matching barrier heights to device requirements.

The passivation of III–V-semiconductors cannot be achieved by hydrogen. The interaction of hydrogen with clean, cleaved GaAs(110) surfaces at low temperatures generates surface states of acceptor character at 0.54 eV above the VBM [30]. This results in a band bending of 0.85 eV on n-type doped samples. Other nonmetallic adsorbates like sulfur, chlorine, bromine, iodine, oxygen, and fluorine also induce surface states of acceptor-type character [26].

Instead of using adatoms to saturate the dangling bonds at the surface, the chalcogen modification described in this work uses a slightly different approach. An exchange reaction between the chalcogen atoms and the group V atoms at the surface results in a thin gallium-chalcogenide like layer at the surface. Since chalcogen atoms have one excess electron compared to the group V atoms, the dangling bonds of chalcogen surface atoms would be double occupied and therefore chemically inreactive. This idea is supported by the results of several experimental investigations [46, 47].

Chalcogen passivation of GaAs(100) surfaces is achieved by wet chemical etching in sulfide solutions. Such surface treatment results in an improvement of the performance of devices like bipolar transistors [48] or laser diodes [49–51]. The chalcogen passivation also improves the structural properties of a wide variety of materials grown on GaAs(100). Here the epitaxial growth of iron films [52] or PTCDA on chalcogen-treated GaAs(100) surfaces should be mentioned [53, 54].

Besides the experimental results presented up to now, the detailed atomic structure is still under discussion. For the Se-passivated GaAs(100) surface, Pashley and Li have proposed a model (4C in Figure 1.8) where the surface is terminated with a layer of selenium dimers, followed by a Ga layer and a second layer of selenium atoms [55, 56]. The fourth atomic layer contains an equal amount of Ga atoms and vacancies and is followed by the GaAs bulk starting with an As layer. This structure model satisfies the electron counting rule [57]. Gundel and Faschinger [58] presented another promising structure model (3B) which has been further supported by DFT-LDA calculations by Benito *et al.* [59]. In this structure model, the Se atoms do not form dimers on the surface and the second Ga layer is free of vacancies. Following the nomenclature introduced by Gundel and Faschinger, the cipher gives the number of Se atoms in the unit cell and the letter discriminates between configurations with the same number of Se atoms.

Two different techniques were used for the chalcogen passivation of GaAs(100): exposure to a flux of chalcogen atoms under UHV conditions and wet chemical etching in sulfur-containing solutions. For the UHV treatment, homoepitaxial n- and p-type GaAs(100) layers with a doping concentration of $N = 1 \times 10^{18}\,\mathrm{cm^{-3}}$ served as substrates. After their growth by molecular beam epitaxy, they were covered by a thick amorphous arsenic layer to protect the GaAs(100) surfaces against contamination and oxidation. These samples were transferred into an ultrahigh vacuum system with a base pressure of $p < 2 \times 10^{-8}$ Pa. The arsenic layer was then removed by gentle annealing to 380 °C. This leads to an As-rich c(4 × 4) or (2 × 4) surface reconstruction of the GaAs(100) surface as can be judged from the line shape analysis of the measured photoemission spectra and additional LEED experiments. The structure models for these As-terminated GaAs(100) sur-

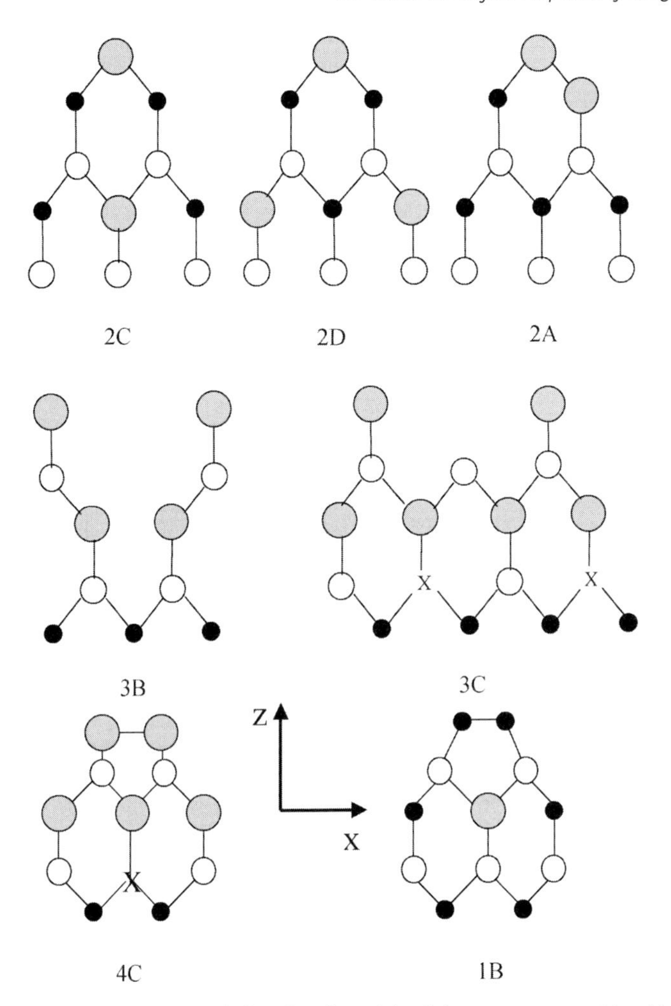

Figure 1.8 Schematic ball-and-stick models of the structures considered in this work for the Se/GaAs(100) system. Gray circles correspond to Se atoms, white circles to Ga atoms, and black circles to As atoms.

faces are shown in Figure 1.9. For the chalcogen passivation, the compounds SnS_2 and $SnSe_2$ were used as source materials. These compounds decompose at 340 °C and 550 °C according to $SnSe_2 \rightarrow SnSe + Se \uparrow$ and $SnS_2 \rightarrow SnS + S \uparrow$, respectively [60]. Sulfur and selenium were evaporated onto the substrates kept at 330 °C and 500 °C, respectively.

For the wet chemical sulfur passivation, samples were first degreased and then etched in a 3:1 mixture of CCl_4 and S_2Cl_2 for 10 s. Rinsing the samples successively in CCl_4, acetone, ethanol, and deionized water for 5 s each followed the etching. After transferring the samples into an UHV system, they are annealed at 430 °C. Both passivation procedures lead to a well-ordered, (2 × 1) reconstructed surface as revealed by LEED [55].

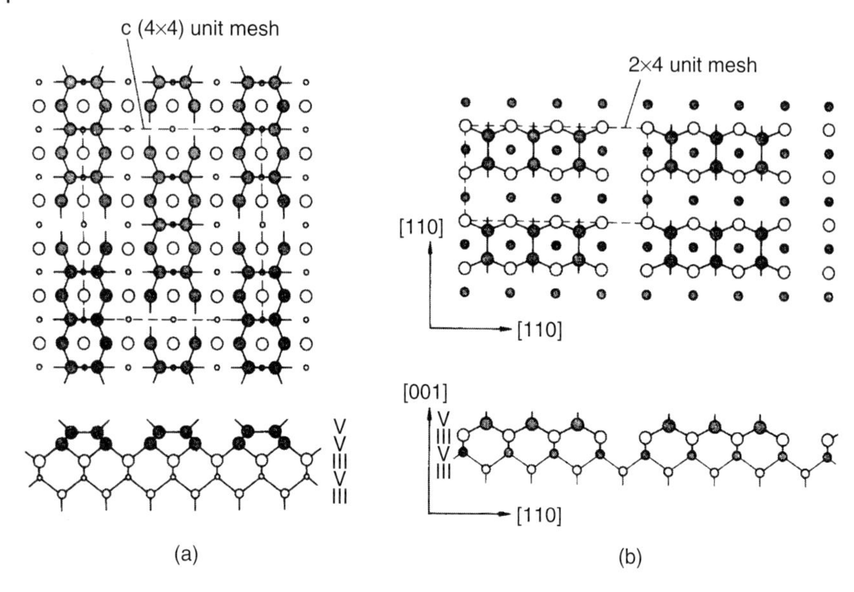

Figure 1.9 Structure model of a (a) GaAs(100)-c(4 × 4) and (b) GaAs(100)-(2 × 4) surface.

Table 1.1 Fit parameters for the Ga3d, As3d, Se3d, and S2p core level spectra.

Core level	Ga3d	As3d	Se3d	S2p
Lorentzian width (eV)	0.1	0.1	0.1	0.1
Branching ratio	1.58	1.50	1.60	2
Spin–orbit splitting (eV)	0.48	0.68	0.86	1.18

The photoemission measurements were performed at the TGM 2 beamline of the synchrotron radiation source BESSY I at Berlin. The UHV chamber at this beamline is equipped with a VG ADES 400 electron spectrometer providing a combined resolution of both light and photoelectrons of about 300 meV at 65 eV photon energy. The photoemission spectra were taken under surface-sensitive conditions, that is, minimum escape depth of the detected photoelectrons.

The photoelectron core level spectra were curve fitted using Voigt profiles – a Lorentzian convoluted with a Gaussian line shape – and a nonlinear least-squares fitting routine. During curve fitting, the Lorentzian linewidth, spin-orbit splitting, and branching ratio were kept fixed at values providing satisfactory results over an entire series of spectra. These parameters are presented in Table 1.1. The peak intensity, position, and Gaussian linewidth were variable. All binding energies are given for the $d_{5/2}$ or $p_{3/2}$ components of the spin–orbit split core levels relative to the Fermi level.

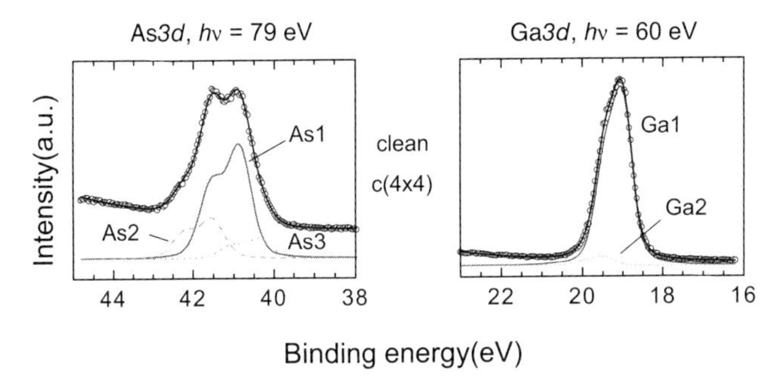

Figure 1.10 As3d and Ga3d core level emission from a GaAs(100)-c(4 × 4) surface.

The As3d and Ga3d core level emission spectra of the GaAs(100)-c(4 × 4) surface are shown in Figure 1.10. The As3d core level spectrum for the clean surface, after decapping the protecting As layer, consists of three components. The As1 component is attributed to As in the fourfold coordinated environment of the GaAs bulk. The components As2 and As3 are shifted by (0.62 ± 0.03) eV and (0.50 ± 0.04) eV toward higher and lower binding energy, respectively. The higher binding energy component originates from As atoms in the surface dimers in the first layer of the As-rich GaAs(100) surface [61, 62]. The As3 component is attributed to threefold coordinated As atoms in the second As layer of the sample. The Ga3d core level is composed of two components: a bulk component (Ga1) and a surface component (Ga2), which is shifted by 0.46 eV toward higher binding energies and is attributed to Ga atoms below the threefold coordinated As atoms.

The As3d, Ga3d, Se3d, and S2p core level emission spectra for the GaAs(100) surfaces are shown in Figure 1.11. For the Se-passivation under UHV conditions, the As3d consists only of one component As1, which is attributed to As in the fourfold coordinated environment of the GaAs bulk. The Ga3d core level consists of two components: a bulk component Ga1 and a surface component Ga2 shifted by 0.37 eV toward higher binding energies. This surface component is attributed to Ga bonded to Se on the surface. The two Se3d components Se1 and Se2 separated by 0.91 eV are attributed to surface and subsurface components, respectively.

The shape of the Se3d core level is similar to the Se3d obtained for Ga_2Se_3, the only difference being a slightly larger energy difference between the Se1 and Se2 components of 1 eV for Ga_2Se_3.

The S-passivated GaAs(100) surfaces obtained by either UHV treatment or wet chemical etching show comparable core level emission spectra. As in the case of Se3d, the S2p core level consists of two components attributed to surface (S1) and subsurface (S2) sulfur. The Ga3d and the As3d are slightly different from the Se-passivated GaAs(100) showing two additional interface components Ga3 and As2. The similarity in the Se3d and the S2p leads to the conclusion that the S-passivation results in the formation of a Ga_2S_3-like layer. The two interface components indicate that the As–S exchange reaction is less efficient than the As–Se exchange reaction, resulting in a less abrupt interface between Ga_2S_3 layer

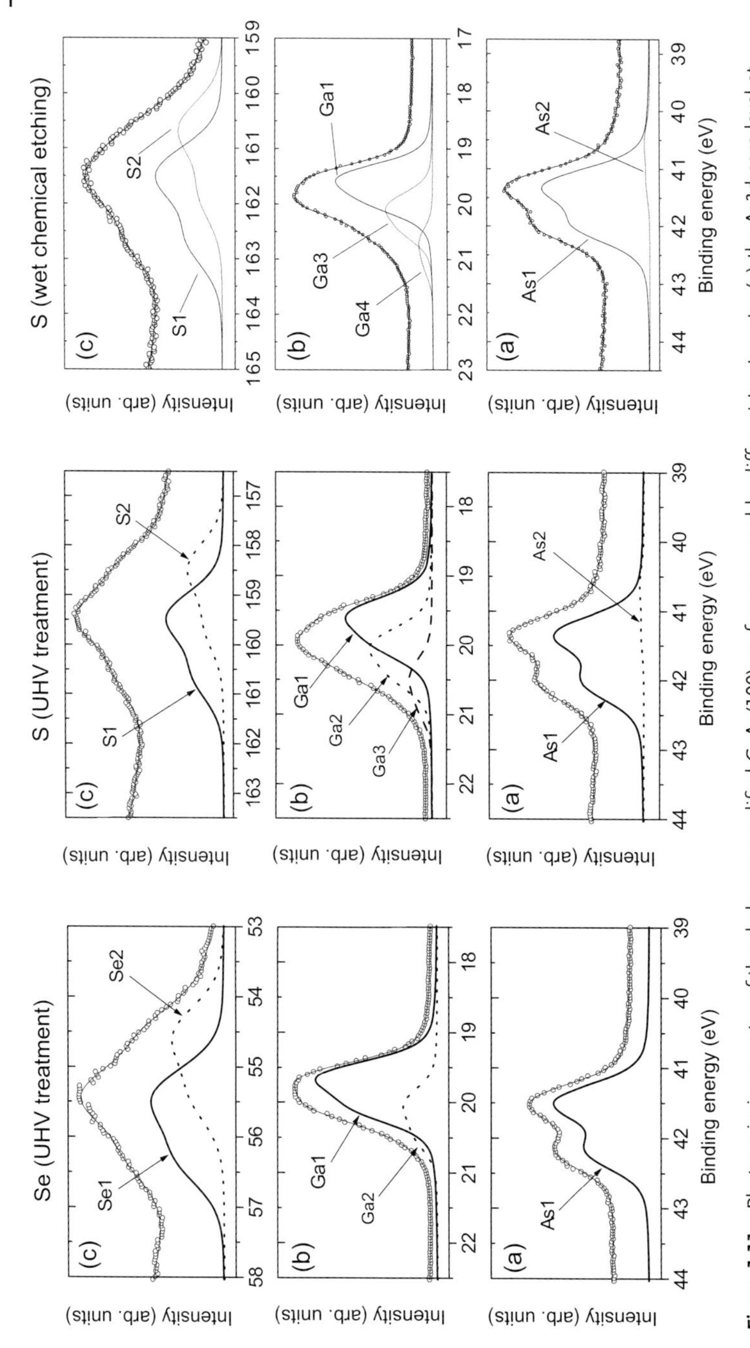

Figure 1.11 Photoemission spectra of the chalcogen-modified GaAs(100) surface prepared by different treatments: (a) the As3d core level at 79 eV photon energy, (b) the Ga3d core level at 60 eV photon energy, and (c) Se3d at 88 eV photon energy/S2p at 195 eV photon energy. Binding energies are given with respect to the Fermi level.

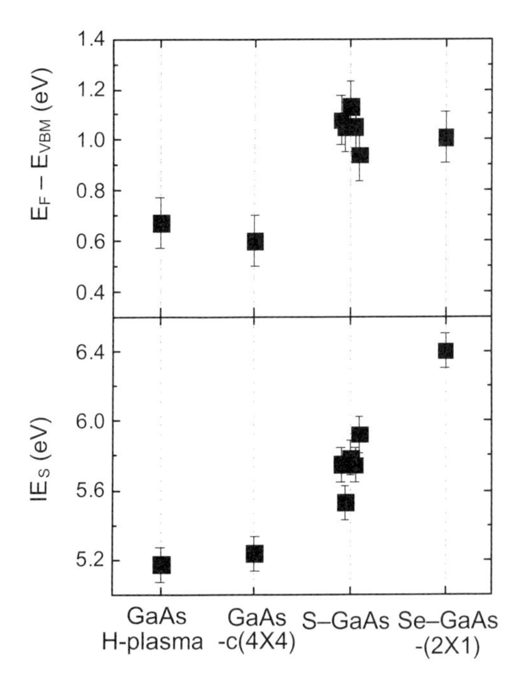

Figure 1.12 Ionization energy IE_S and position of the Fermi level with respect to the valence band maximum $E_F - E_{VBM}$ for differently treated GaAs(100) surfaces.

and GaAs bulk. This is supported by the fact that a higher temperature is necessary for the S-passivation. All passivation processes result in surfaces showing a (2×1) reconstruction, which survives considerable exposure to air, revealing the chemical stability of the passivated surfaces. Since the As3d core level shows only one component and the chalcogen atoms are found in two different chemical environments, the number of possible structure models is reduced to structure 3B, 3C, and 4C.

The ionization energy and the position of the Fermi level with respect to the valence band maximum on the chalcogen-passivated surfaces compared to nonpassivated surfaces are shown in Figure 1.12. The Fermi level is at about 0.65 and 0.6 eV above the valence band maximum on samples cleaned by an H-plasma or by decapping of the As layer, respectively. The chalcogen treatments shift the Fermi level by about 0.4 eV toward the conduction band minimum. Compared to nonpassivated samples, the band bending is thus reduced to 0.4 eV for n-type GaAs. It should be mentioned that the energy position of the Fermi level on p-type samples is almost independent of the sample treatment and amounts to about (0.5 ± 0.1) eV above the valence band maximum. Therefore, it can be concluded that the chalcogen passivation does not result in an electronic passivation, that is, the surfaces still show band bending due to surface states of acceptor as well as donor-type character.

The ionization energy increases as a function of the chalcogen treatment. Since S and Se have a larger electronegativity than Ga, negative charge is transferred to the chalcogen atoms. This results in a surface dipole, which increases the ionization energy. The larger change in ionization energy due to the Se treatment may be explained by the more efficient formation of a Ga-chalcogenide-like layer.

The maximum change in ionization energy will now be calculated using Eqs. (1.9)–(1.11). With the Pauling electronegativities of $X_{ga} = 1.81$ and $X_S = 2.58$, the charge transfer between Ga and S amounts to $\Delta q = 0.144$. The distance between the S and Ga in the first surface layer perpendicular to the surface amounts to 0.11 nm [63]. These two values give a dipole moment normal to the surface of 2.54×10^{-30} cm. The density of surface dipoles is assumed to be equal to the density of atoms on a GaAs(100) of $6.26 \times 10^{14}\,\text{cm}^{-2}$ (see Reference [26]). With the polarizability for S of $2.9 \times 10^{-24}\,\text{cm}^{-3}$, the change in ionization energy amounts to 1.28 eV. The same calculation can be done for the Se-covered surface using $2.9 \times 10^{-24}\,\text{cm}^{-3}$ and 2.55 for the polarizability and electronegativity of Se, respectively. Here, the ionization energy is expected to vary by 1.12 eV. Both chalcogen atoms are negatively charged due to their higher electronegativity compared to Ga and the ionization energy is expected to increase. Direction and magnitude of the change in ionization energy agree perfectly with the experimental results shown in Figure 1.12.

The three structure models 3B, 3C, and 4C (see Figure 1.8) supported by the photoemission spectroscopy data will now be investigated in more detail by energy minimization calculations using a first-principles local-orbital code (Fireball96) [64]. Here, structure 3C is by 4.0 eV energetically less favorable than the structure 3B due to creating the fourth-layer Ga vacancy. This leaves the structures 4C and 3B where the latter one is the most stable one. For these two structures and structure 1B, which represents the arrangement of As atoms on the clean substrate surface after As decapping, the corrugation in the STM topography along the directions defined by the dimers in 4C and 1B is calculated. The STM currents between a tungsten tip and the substrate are obtained using a LCAO method based on a local-orbital LDA calculation. The respective line scans for $V = 3.0$ eV and $I = 0.1$ nA are presented in Figure 1.13. The corrugation for structure 3B is 0.5 Å and corresponds well with the experimentally determined corrugation of 0.7 Å obtained by Pashley and Li [55]. The corrugation of the structure 4C proposed by Pashley and Li is less than 0.2 Å and smaller than the theoretical corrugation for structure 1B. In conclusion, the theoretical STM results support the 3B model as the microscopic structure for the Se-passivated surface with only a single Se atom in the topmost layer.

Figure 1.14 shows the electronic band structure of the relaxed 3B geometry. The energy gap amounts to ~2 eV and is very close to the energy gap calculated for bulk GaAs. It should be noted that in minimal basis set calculations, the conduction band is not very well reproduced, giving an overestimation of a few tens of an electron volt. The topmost valence band is associated with the surface Se atom, while the bottommost conduction band is associated with the bonds formed between the Se and Ga atoms of the third and fourth layer. The topmost valence band has a p_x character and shows a small dispersion, which is the result of the

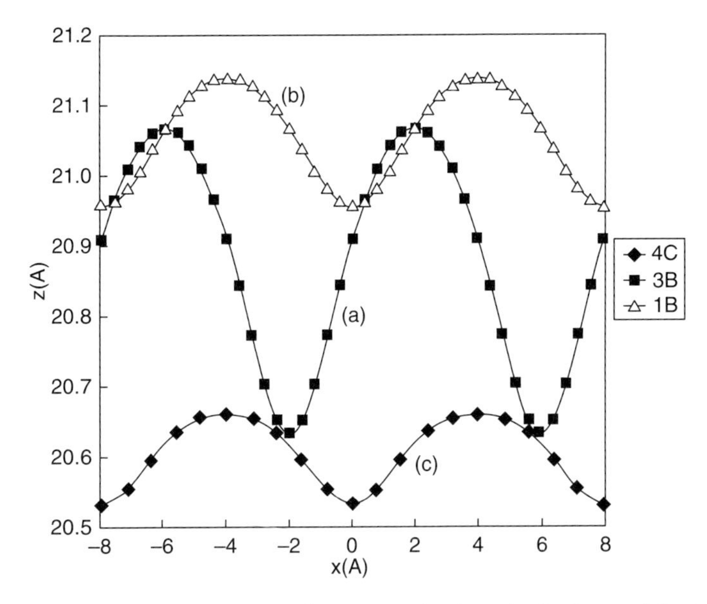

Figure 1.13 Corrugation in surface topography along the Se dimers of the structure 4C (c) compared to structures 3B (a) and 1B (b) in the same direction. Units are given in Å.

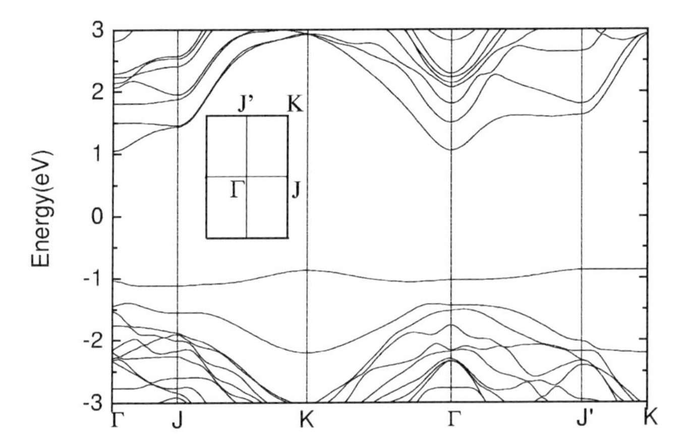

Figure 1.14 Surface bands for structure 3B in Figure 1.8. The zero of energy corresponds to a position close to the middle of the GaAs semiconductor gap.

large distances, around 4.0 Å, between the Se nearest neighbors in the y-direction and the π-character of the interaction between these p_x orbitals.

In contrast to the theoretical predictions that the band gap of these surfaces is free of states, the sample prepared by the procedures described here still exhibit surface states. These surface states may be attributed to defects or dopant atoms at surface.

Since UPS and LEED give almost identical results for the Se- and S-passivated GaAs(100) surfaces, the structure model developed for the Se-passivated will also be applied to the S-passivated surface.

1.2 Semiconductor Interfaces

Most metal–semiconductor contacts are rectifying [65]. Schottky explained this behavior by depletion layers on the semiconductor side of such interfaces [66]. The band bending in this space charge region is characterized by its barrier height, which is the energy distance between the Fermi level and the edge of the respective majority-carrier band at the interface. In the first approaches to describe the band line-up in metal–semiconductor contacts, only charge carrier transport over the barrier was considered and interface states were not taken into account. In this simple picture, the vacuum levels of the metal and the semiconductor are aligned at the interface, that is, no interface dipole exists. In this case, the barrier height is found to be the difference between the workfunction of the metal and the electron affinity of the semiconductor. This is the famous Schottky–Mott rule [67, 68]. For a given semiconductor, the barrier heights of different metal contacts should scale linearly with the workfunction of the metal and the slope should be unity. Applying the same model to semiconductor heterostructures, the energy difference in the conduction band minima at the interface is given by the difference of the electron affinities of the semiconductors (*Anderson rule*) [69].

Already, Schottky presented in his famous paper published in 1940 data from Schweikert [70], which clearly display that the Schottky–Mott rule does not hold. The barrier heights of metal–selenium contacts are found to scale with the workfunction of the metal, but the slope is found to be smaller than unity. Later on, Bardeen proposed that interface states are responsible for the shortcomings of the Schottky–Mott rule [71].

1.2.1 Metal–Semiconductor Contacts

The electric transport across metal–semiconductor contacts is carried by the majority carriers and is thus characterized by their barrier heights. There are several physical mechanisms that may determine the barrier heights of Schottky contacts. Ideal interfaces are intimate, abrupt, laterally homogeneous, and free of structural defects as well as foreign atoms. Their barrier heights are then determined by the continuum of metal-induced gap states (MIGSs), or more generally, interface-induced gap states (IFIGSs) [28]. These intrinsic interface states derive from the ViGS of the complex semiconductor band structure, which have been introduced in Section 3.1. They represent the wave-function tails of the metal electrons into the semiconductor in the energy range between the valence-band maximum and the Fermi level where the metal conduction band overlaps the semiconductor band gap. At real but still intimate and abrupt contacts, interface dipoles induced by foreign atoms or by specific interface structures as well as structural and chemical interface point defects may be present and will then contribute to the barrier heights as secondary mechanisms in addition to the MIGS (see, e.g., References [72] and [73]). Strongly intermixed metal–semiconductor contacts are even more complicated and their barrier heights are beyond description by simple models. The MIGSs derive from the bulk bands and, thus, their character changes across the band gap from more acceptor-like closer to the conduction band to predominantly donor-like nearer to the valence band. When the respective branch point is above, coincides with, or drops below the Fermi level, the net charge in the MIGS continuum has positive sign, vanishes, and becomes negative, respectively. Thus, these branch points have the significance of charge neutrality levels (CNLs). Interfaces are electrically neutral. Therefore, the charge in the MIGS continuum is compensated by an equal amount of charge but of opposite sign on the metal side of the interface. The space charge on the semiconductor side may be neglected since it is small compared to what is found in the MIGS. A chemical approach would assign such charge transfer at metal–semiconductor contacts to the ionicity of the heteropolar bonds between metal and substrate atoms at the interface. As shown in Section 3.2, Pauling correlated the ionic character of single bonds in diatomic molecules with the difference of the atomic electronegativities [34]. A generalization of this chemical concept then describes the charge transfer across metal–semiconductor interfaces by the difference of the metal and the semiconductor electronegativities. The combination of the physical MIGS and the chemical electronegativity concepts predicts the barrier heights to vary as [74]

$$\phi_{\mathrm{Bn}} = \phi_{\mathrm{CNL}} + S_X(X_M - X_S) \tag{1.12}$$

and

$$\phi_{\mathrm{Bp}} = \phi^*_{\mathrm{CNL}} - S_X(X_M - X_S) \tag{1.13}$$

on semiconductors doped n- and p-type, respectively. The zero charge barrier heights $\phi_{CNL} = W_C - W_{bp}$ and $\phi^*_{\mathrm{CNL}} = W_{\mathrm{bp}} - W_V$ result when the difference $X_M - X_S$ between the metal and the semiconductor electronegativities is zero and, consequently, the Fermi level W_{F} coincides with the branch point W_{bp} of the MIGS. In this special case, the vacuum levels align at the interface and a situation as predicted by the Schottky–Mott rule is reached. The energies of the conduction- and the valence-band edges are denoted as W_C and W_V, respectively.

The MIGS-and-electronegativity model describes the barrier heights of ideal metal–semiconductor contacts by two parameters, which are the zero-charge-transfer barrier height, and the slope parameter $S_X = \partial\phi_B/\partial X_M$. A semiempirical rule [75] that was later justified theoretically [76] relates the slope parameters with the optical dielectric constant ε_∞ of the semiconductors as

$$A_X/S_X - 1 = 0.1(\varepsilon_\infty - 1)^2 \tag{1.14}$$

The parameter A_X depends on the electronegativity scale used and one obtains $A_X = 0.86$ for Miedema's solid-state electronegativities [77]. Tersoff calculated the energy positions $W_{bp} - W_V$ of the branch points with regard to the valence-band maxima for 15 semiconductors [78, 79]. He obtained the energy bands with a linearized augmented planewave method and the local-density approximation for correlation and exchange, and adjusted the underestimated band gaps to their experimental values by applying the "scissors operation."

Recently, Mönch used a computationally much simpler approach [80]. He combined Baldereschi's concept [81] of mean-value points in the Brillouin zone and Penn's idea [82] of the average or dielectric band gap of semiconductors. By a comparison with band structures computed by Rohlfing *et al.* in the GW approximation for six semiconductors [83, 84], Mönch demonstrated that the dispersion of the valence bands is well approximated by the empirical tight-binding method (ETB) and the widths of the band gaps at the mean-value point equal to the dielectric band gaps W_{dg}. Considering Tersoff's data mentioned earlier, he arrived at the energy positions [80]

$$W_{bp} - W_v = 0.446\, W_{dg} - [W_V - W_V(k_{mv})]_{ETB} \tag{1.15}$$

of the branch points above the valence-band maximum W_V in the middle of the Brillouin zone. The energy differences $[W_V - W_V(k_{mv})]$ of the valence band at the Γ and at the mean-value point k_{mv} are ETB values. For GaAs, the branch point energies for p- and n-type doping amount to 0.52 and 0.9 eV, respectively. With $\varepsilon_\infty = 10.9$ for GaAs (1.14) gives a slope parameter of 0.08 for GaAs.

The concept of interface states also applies to semiconductor–semiconductor interfaces. Here, for the situation displayed in Figure 1.15, the valence band and conduction band of the semiconductor on the left partly overlap with the band gap of the semiconductor on the right. This will result in interface states within the

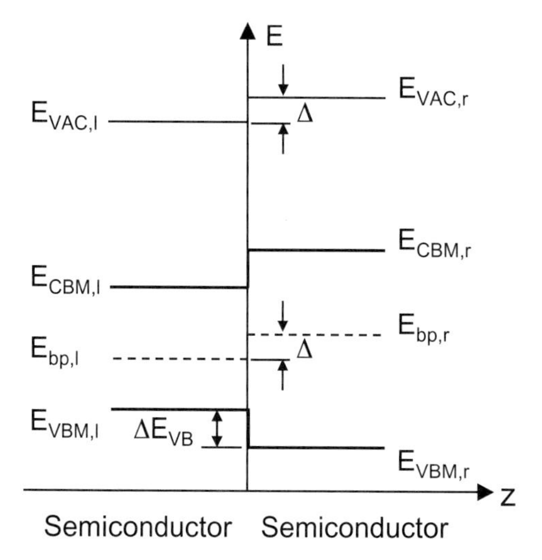

Figure 1.15 Band structure of a semiconductor heterostructure with an interface dipole.

band gap of the semiconductor on the right. For abrupt, defect-free semiconductor–semiconductor interfaces with no net charge transfer across the interface, the band should align in such a way that the charge neutrality levels of the two semiconductors in contact align. An additional charge transfer across the interface will result in an interface dipole and contribute to the energy level alignment as shown in Figure 1.15.

1.2.2
Metal Contacts on GaAs

Figure 1.16 shows that the experimentally determined barrier heights follow very well the trend predicted by the MIGS-and-electronegativity model. The barrier

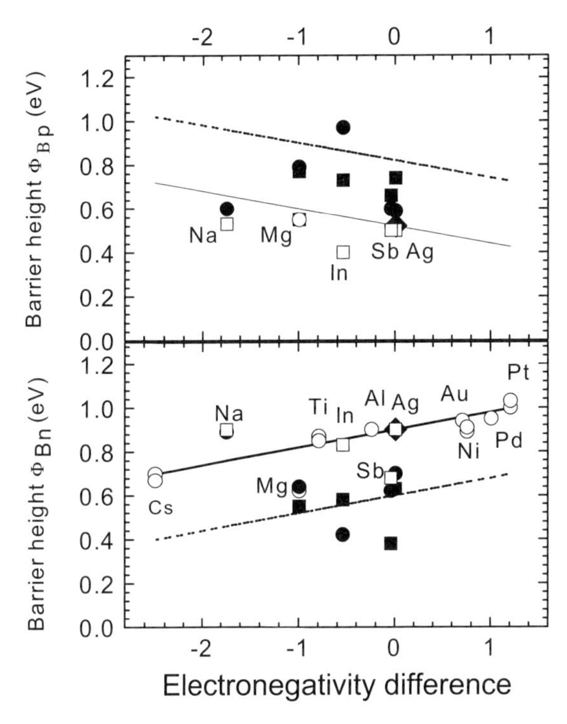

Figure 1.16 Barrier heights of laterally homogenous GaAs Schottky contacts versus the difference in Miedema electronegativity of the metal and the GaAs. *IV* results (○): Ag:Hardiker *et al.* [91]; Al: Bhuiyan *et al.* [92]; Ti: DiDio *et al.* [93] Arulkumaran *et al.* [94]; Nuhoglu *et al.* [95]; Ni: Hackham & Harrop [96]; Nathan *et al.* [97]; Pd: Dharmarusu *et al.* [98]; Pt: Barnard *et al.* [99]; Hübers & Röser [100]; Mg: Waldrop [101]; Photoemission spectroscopy results (□); Cs: Spicer *et al.* [102]; Grunwald *et al.* [103]; Ag: Vitomirov *et al.* [62]; Sb: Cao *et al.* [90]; In: Mao *et al.* [104]; Na: Prietsch *et al.* [105]; Se-modified contacts (●); Ag: Hohenecker *et al.* [106]; Mg: Hohenecker *et al.* [107]; Sb: Hohenecker *et al.* [107]; In: Hohenecker *et al.* [108]; Na: Hohenecker *et al.* [291]; S-modified contacts (■): Ag: Hohenecker *et al.* [109]; Mg: Hohenecker *et al.* [43]; Sb: Hohenecker *et al.* [110]; In: Hohenecker *et al.* [44]. The straight line represents the prediction of the MIGS theory and ◆ the CNL.

heights presented here are obtained from current voltage (*IV*) and photoemission measurements on samples where a special emphasis was laid on preparing abrupt and defect-free interfaces. It is a common procedure to determine barrier heights of metal–semiconductor contacts from their *IV* measurements. Tung *et al.* [85, 86] and Rau *et al.* [87] already pointed out that inhomogeneities might play an important role and have to be considered in the evaluation of the experimental *IV* characteristics. The application of standard procedures considering thermionic emission of charge carriers gives *effective* barrier heights and ideality factors only. Both parameters vary from diode to diode even if they are identically prepared. A correlation between effective barrier heights and ideality factors was reported and approximated by a linear relationship [88]. This finding was attributed to inhomogeneous interfaces, and the barrier heights obtained by extrapolation to the ideality factor calculated for image force lowering only were taken as values characteristic of homogenous interfaces. It was concluded that these values, rather than mean values obtained from the set of identical prepared contacts of the same kind, should be compared with theoretical results. Accordingly, barrier heights determined from *IV* measurements presented in Figure 1.16 have been obtained by such a procedure. There is an excellent agreement between the experimental results and the prediction of the MIGS-and-electronegativity model. The barrier heights of Ni and Pd contacts are found to be below the MIGS line due to the fact that both transition metals decompose GaAs at room temperature. In these cases, nonabrupt interfaces are formed.

In comparison to the barrier heights of metal contacts on bare GaAs surfaces, Figure 1.16 shows the barrier heights for metal contacts prepared on S- or Se-passivated GaAs(100) surfaces. First, the chemical properties of these metal-chalcogen-passivated-GaAs interfaces will be discussed. Depositing Mg on Se- or S-modified GaAs surfaces results in an exchange between Mg and Ga atoms, leading to the formation of Mg chalcogenides and Ga clusters, the latter one segregating on the surface. In addition, Mg reacts with the GaAs bulk resulting in the formation of Mg–As compounds that also segregate to the surface. As a result, the interface between the reacted surface layers and the GaAs bulk is shifted into the direction of the GaAs bulk as a function of Mg coverage. Similar observations have been made for the deposition of Na. In both cases, a Fermi edge is observed at sufficient high coverages indicating a metallization of the surface.

The deposition of In, Ag, and Sb leaves the surfaces of the Se- or S-modified GaAs nearly unaffected. The interaction between these metals and the substrate surfaces is limited to the bonding of the metal atoms to the chalcogen surface atoms. For In and Ag, the formation of islands is observed for coverages above 0.24 and 0.12 nm, respectively. For both metals, the clustering is stronger on S than on Se-modified GaAs surfaces, which is attributed to the fact that the S-passivation is less efficient than the Se-passivation, resulting in a less abrupt interface between the chalcogenide-like surface layer and the GaAs bulk. The growth modes of Ag and In on the chalcogen-modified surfaces agree with the respective growth modes on unmodified surfaces. A Fermi edge is observed on Ag and In layers at 0.48 and 0.60 nm nominal thickness, respectively. For Sb, islanding is observed for a nominal coverage above 0.17 nm. This growth mode is in contrast to the layer-by-layer growth mode, which was observed for the Sb deposition on clean and unmodified GaAs(100) surfaces. Sb layer does not show any Fermi edge up to the maximum thickness of 3.3 nm, which shows that it is semiconducting.

For the chalcogen-modified metal/GaAs(100) contacts, the barrier heights were determined by measuring the position of the Fermi level at different modification steps. At high metal coverages, the metal conduction band obscures the valence band edge of the underlying semiconductor. Here, the Fermi level was measured with respect to the bulk components of the substrate core level emissions.

For the same metal–semiconductor contact, the barrier heights for n- and p-type doping of the semiconductor should add up to the band gap of the semiconductor. Comparing the Fermi level positions for the different metals evaporated on chalcogen-modified on n- and p-type doped GaAs samples, it is found that they follow this rule quite well. Exceptions are Sb contacts, which is due to the fact that the Sb films are nonmetallic. For Sb coverages above 100 monolayers, which is much higher than what has been deposited in this investigation, Sb layers grown on n- and p-type doped GaAs(110) show a metallic behavior [89]. The saturation value for the Fermi level on Se- and S-modified n-type (p-type) GaAs(100) amounts to 0.81 eV (0.60 eV) and 1.05 eV (0.66 eV), respectively. This is considerably higher than Fermi level positions for Sb on unmodified n- and p-type doped GaAs(110), which amount to 0.75 and 0.5 eV, respectively [90].

The barrier heights of Schottky contacts on S- and Se-passivated GaAs(100) show a general trend. They are larger and smaller than the barrier heights on nonpassivated p- and n-type doped GaAs(100) substrates, respectively. This trend can be explained within the MIGS-and-electronegativity model assuming that the chalcogen-induced surface dipole still exists at the metal–semiconductor interface and contributes to the charge transfer across the interface. The electronegativity of the passivating atoms is larger than the electronegativities of the substrate atoms. This results in charge of positive sign on the semiconductor side of the interface, which is compensated by equal charge of opposite sign in the MIGS. Therefore, the Fermi level moves closer to the CBM, that is, the barrier heights on n- and p-type substrates are decreased and increased, respectively. The change in barrier height can be estimated from Eq. (1.11) by

$$\Delta\phi_{\mathrm{Bi}} = \Delta I/\varepsilon_I \tag{1.16}$$

where ε_I is the dielectric constant of the semiconductor at the interface [1, 16]. With a dielectric interface constant of 4 for GaAs (see Reference [26]), the S- and Se-induced change in barrier height amounts to 0.32 and 0.28 eV, respectively. This change is indicated by the dashed lines in Figure 1.16, which are shifted by 0.3 eV with respect to the MIGS theory line.

The chalcogen modification may also result in change in the position of the CNL. Theoretical values for the CNL on such surfaces are not available. However, a CNL can be determined from the energy level alignment at PTCDA/S–GaAs(100) interfaces, as will be shown in Section 6.3.

2
Growth of Thin Films

Thin films of organic semiconductors are prepared using similar techniques as for the deposition of inorganic semiconductors or metals. Since most molecular semiconductors have a low solubility in liquids, liquid-phase thin film deposition techniques like spin coating or Langmuir–Blodgett are not the techniques of choice. Therefore, the two techniques most often used are organic molecular beam deposition (OMBD) and organic vapor phase deposition (OVPD); the later technique is mostly used for the growth of devices and commercial applications. A prerequisite of both techniques is that the source material is chemically clean. It should be mentioned that some materials used for devices are offered by suppliers in a purified version.

Compared to the deposition of atoms, OMBD has its own particular challenges. While crossing the melting point of atomic materials just increases the vapor pressure/evaporation rate, this would destroy most organic materials. Therefore, organic semiconductor deposition is mostly done by sublimation. In addition, molecules are large entities compared to atoms, that is, they already have a certain, mostly anisotropic structure. This has implications for the thin film formation, that is, change of orientation and/or conformation as a function of film thickness, or the formation of domains. These processes are enabled by the usually weak interactions of molecule–substrate and molecule–molecule.

Figure 2.1 shows there different growth mode scenarios, which are also observed during the growth of organic semiconductor thin film. In a phenomenological approach, these different growth modes can be explained by taking into account the surface energies of the substrate (γ_{sub}), the film (γ_{film}), and the interface between film and substrate (γ_{int}). In the case of island and layer-by-layer growth, the sum of film and interface surface energy $\gamma_{film} + \gamma_{int}$ is larger or smaller than the substrate surface energy γ_{sub}, respectively. During layer-plus-island growth, the conditions change as a function of film thickness.

Low Molecular Weight Organic Semiconductors. Thorsten U. Kampen

ISBN: 978-3-527-40653-1

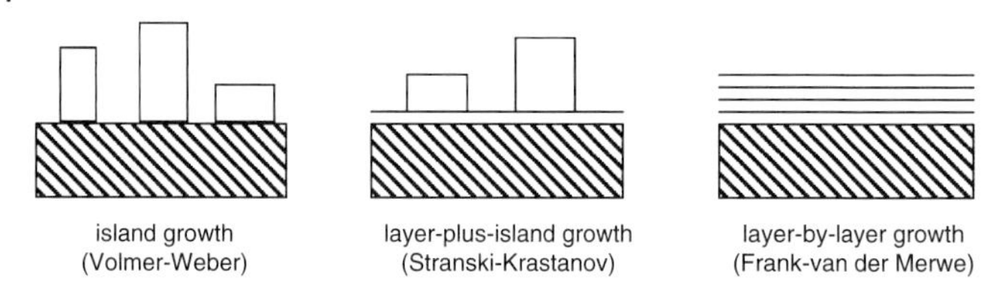

Figure 2.1 Schematic representation of the three important growth modes.

2.1 Purification

The density of atoms in inorganic semiconductors is of the order of $10^{23}\,cm^{-3}$. Intentional impurity concentration (doping) of 10^{15}–$10^{19}\,cm^{-3}$ in inorganic semiconductors changes the physical properties, like charge transport and optical absorption or emission. In silicon, for example, doping concentrations $10^{14}\,cm^{-3}$ and $10^{19}\,cm^{-3}$ shift the Fermi level form the center of the band gap close to one of the band edges, changing the semiconductor form being intrinsic toward degenerated. This changes the charge transport properties as well as the properties of electrical contacts. Non-intentional impurities, on the other hand, will drastically decrease the thin film properties. They will scatter or trap electrons, or serve as non-radiative recombination centers. The same argument also holds for organic semiconductors and, therefore, purification prior to thin film deposition becomes mandatory.

In organic materials, there are several sources of impurities. On one hand, impurities can be remaining molecules involved in the synthesis of the material. On the other hand, impurities can be created by the decomposition of the molecule at higher temperatures during the evaporation process. For PTCDA, decarboxylation takes place at a temperature of about 720 K [111]. Mass spectrometry investigations on nonpurified PTCDA show that the decomposition is enhanced by the presence of water leading to a hydrolization of the anhydride group [112]. This hydrolization is observed at temperature of 470 K, which is well below the sublimation temperature. Figure 2.2 shows a mass spectrum (b) taken during the evaporation of PTCDA, while spectrum (a) represents the mass spectrum of the residual gas in the UHV system with the PTCDA evaporator being switched off. The peak at 392 amu is due to the PTCDA cation, while peaks at lower amu are due to fragments. The water in the material is incorporated in the small crystallites of the source material. Even extended degassing in UHV is not sufficient to remove this water. As will be shown later, these impurities may be observed in photoemission spectroscopy as well as in charge transport measurements.

The methods used for purification rely on the fact that vapor pressures are different for different molecules. The source material is heated in a glass tube under vacuum. This glass tube is placed in a furnace, which has a temperature gradient as shown in Figure 2.3.

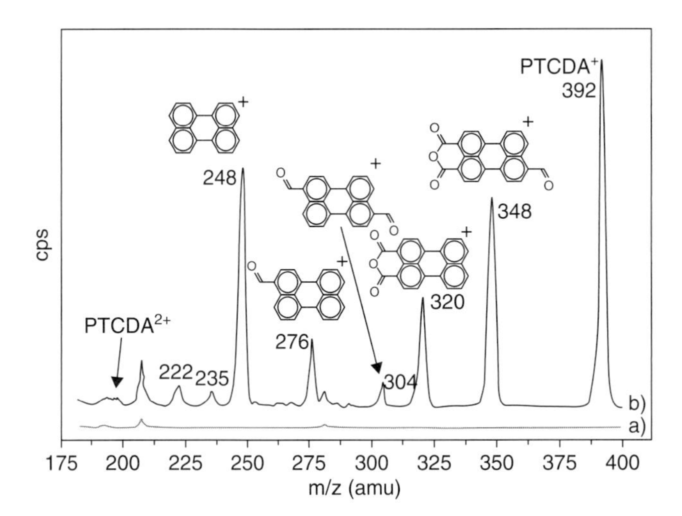

Figure 2.2 (a) Background spectrum. (b) Possible fragments of the PTCDA molecule as observed with mass spectrometry during evaporation. The peaks at 222 and 235 amu show the same temperature-dependent intensity behavior as the PTCDA peak and can thereby be assigned also to be PTCDA fragments (from Reference [112]).

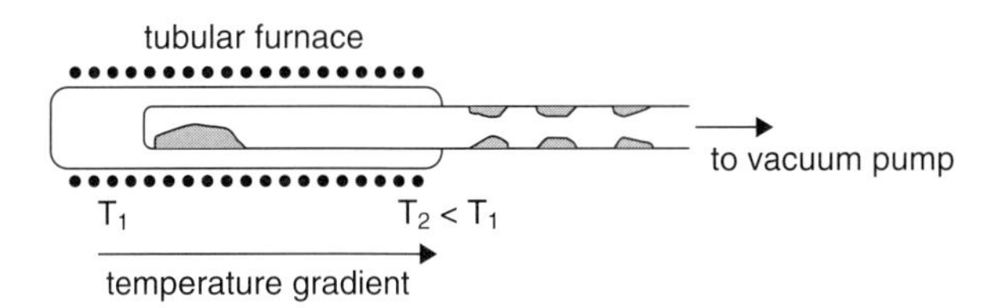

Figure 2.3 Temperature gradient sublimation setup for purification of organic materials.

The different components of the sublimated source material condense at different temperatures and, hence, places in the glass tube. The procedure may be repeated with the purest fraction as the new starting material. Instead under vacuum, this procedure may also be performed using a weak stream of Ar or N_2. The gas then drags along the molecules. This principle is used on a larger scale in OVPD as will be explained in Section 2.3.

In addition, materials may also by purified by chromatographic methods, but a prerequisite of this method is the solubility of the source material in a solvent, which seems to be a limitation for small molecules. For example, PTCDA is found to be weakly soluble in dimethyl sulfoxide, $(CH_3)2S{:}O$ (DMSO), with concentrations limited to only $2 \pm 1\,\mu M$ [13].

After purification, some molecules have to be kept in a protecting atmosphere of an inert gas to keep them from oxidation. Pentacene, for example, is found to oxidize easily in air [113].

2.2 Organic Molecular Beam Deposition

OMBD is the most often used technique to deposit thins films of small molecules on substrates. Here, the more general term *deposition* is used to include systems where there is no epitaxial relation to the substrate. In OMBD, the molecules are thermally evaporated from evaporation cells. For a reasonable evaporation rate, the vapor pressure should be around 10^{-4} mbar. The evaporation rates can be monitored by a quartz crystal microbalance or a mass spectrometer. Most organic materials reach this vapor pressure at much lower temperatures compared to inorganic materials. Typical evaporation temperature is up to 400 °C, and temperature control is much more critical compared to inorganic materials as they are destroyed easily by using too high temperatures. Therefore, evaporation is done at temperatures below the melting point by sublimation. Since UHV systems are usually baked at 120–200 °C, the mounting of the evaporation cells of a given organic materials may depend on its vapor pressure at the bake-out temperature. If the vapor pressure is still low, the evaporator may be directly mounted to the UHV chamber (Figure 2.4). If, on the other hand, the evaporation temperature is lower than the bake-out temperature, the evaporator should be mounted to the deposition chamber via a gate valve and a z-translator. Therefore, the evaporator can be mounted after bake-out. After mounting the evaporator, the small volume may be evacuated via a bypass to a turbo pump. The gate valve may be opened and the evaporator moved into the deposition chamber. Mounting evaporators via gate valves has some further advantages:

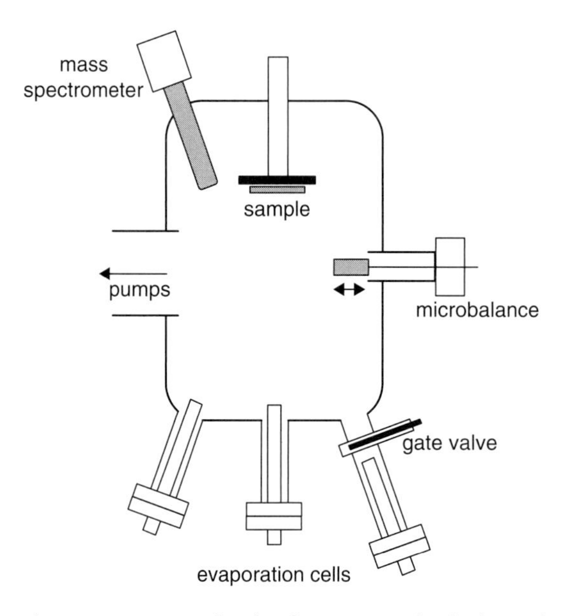

Figure 2.4 Vacuum chamber for organic molecular beam deposition.

- The source material or the evaporator may be exchange without venting the whole deposition chamber.
- The source material may be carefully heated for degassing without exposing the deposition chamber.
- A residual gas analyzer may be attached to the bypass for controlling the cleanliness.

2.3 Organic Vapor Phase Deposition

The principle of OVPD is shown in Figure 2.5. The source material is heated in crucible and thermally evaporated into a stream of an inert carrier gas. The sublimated molecules are carried by the inert gas toward the cooled substrate. The gas flow is widened in the vicinity of the substrate. The whole growth chamber is heated so that condensation of the molecules on the chamber walls is prevented. The evaporation of the source material is technically separated from the deposition chamber, and the deposition rate is controlled by the gas flow using mass flow controllers. This also enables doping by mixing gas flows of different molecules before they enter the deposition chamber. This technique works at pressures around 0.5–1 mbar. It can be used for the precise deposition on large substrates. Compared to OMBD, all evaporated source material is efficiently used for the thin film deposition, which is important because some organic semiconductors are very expensive. Many features of OVPD make it suitable to industrial application and this technique, developed by Stephen Forrest in Princeton, USA, is now commercialized by AIXTRON in Aachen, Germany. It has been successfully used for the manufacturing of devices, for example, pentacene devices and logic gates [114], photovoltaic cells [115].

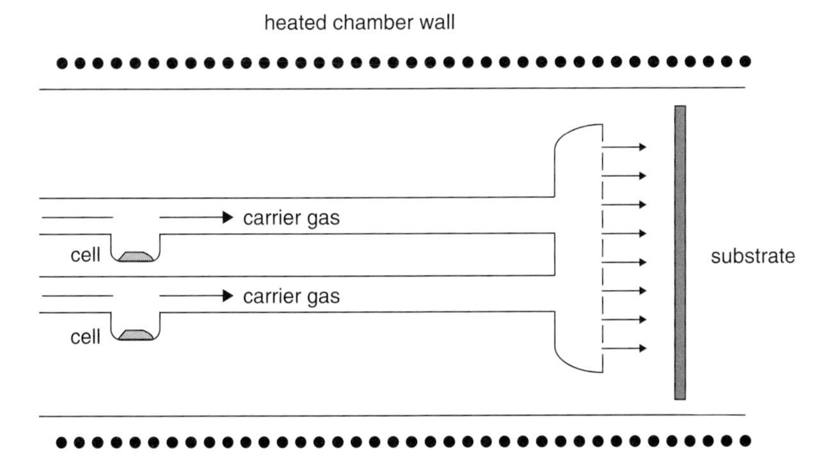

Figure 2.5 Principle of organic vapor phase deposition.

3
Structural Analysis

Both perylene derivatives form monoclinic crystalline structures (space group $P2_1/c(C^5_{2h})$). For PTCDA, two different crystalline modifications called α- and β-phase are known [116–118]. Figure 3.1 illustrates how the molecules in a PTCDA crystal form stacks, which are parallel to the *a*-direction in the crystal. This direction is also the direction of growth of the needle-like PTCDA crystallites. The equal orientation of molecules within all stacks results in a layered structure with the molecules being parallel with their molecular plane with respect to the (102) lattice plane. The molecules are not absolutely parallel to the (102) plane but tilted by an angle of 11° around their long symmetry axis [119, 120].

The distance between next-neighbor (102) planes amounts to 0.322 nm and is smaller than the next-neighbor distance of 0.335 nm for the planes in graphite. Therefore, the overlap of the highest occupied molecular orbitals (HOMOs), which consists of π-orbitals, is larger in PTCDA and results in strong anisotropic optical and electrical properties. The lattice parameters of both modifications given in Figure 3.1 differ only slightly. The main difference in both modification is due to the fact that in the α- and β-modification, the kursio a-direction is tilted with respect to the *c*- and *b*-axis, respectively. This reduces the overlap between molecules in adjacent (102) crystalline planes. Both polymorphs are found in thin films. Figure 3.1 also presents the unit cell, which contains two molecules. The volumes for both modification differ slightly resulting in different densities of 1.24 and 1.22 g/cm^{-3} for the α- and β-modification, respectively.

For DiMe-PTCDI, only one crystalline structure has been found which is shown in Figure 3.2. As with PTCDA, the crystalline structure is monoclinic with two molecules per unit cell. Due to the methyl groups, the shape of the molecule is more elliptical compared to the almost rectangular shape of PTCDA. This results in a smaller angle of 37° between the long symmetry axes of two molecules in a unit cell [121].

The structural properties of PTCDA and DiMe-PTCDI films depend strongly on the substrate surfaces used for deposition. On substrates like Ag(111) [122], Au(111) [123], and HOPG [124], PTCDA is adsorbed in a "herring bone" structure with two molecules in each unit cell almost perpendicular to each other and parallel to the substrate surface. This arrangement of the molecules is similar to the one in the (102) plane of a PTCDA crystal. On Ag(110), a phase transition is

Low Molecular Weight Organic Semiconductors. Thorsten U. Kampen

ISBN: 978-3-527-40653-1

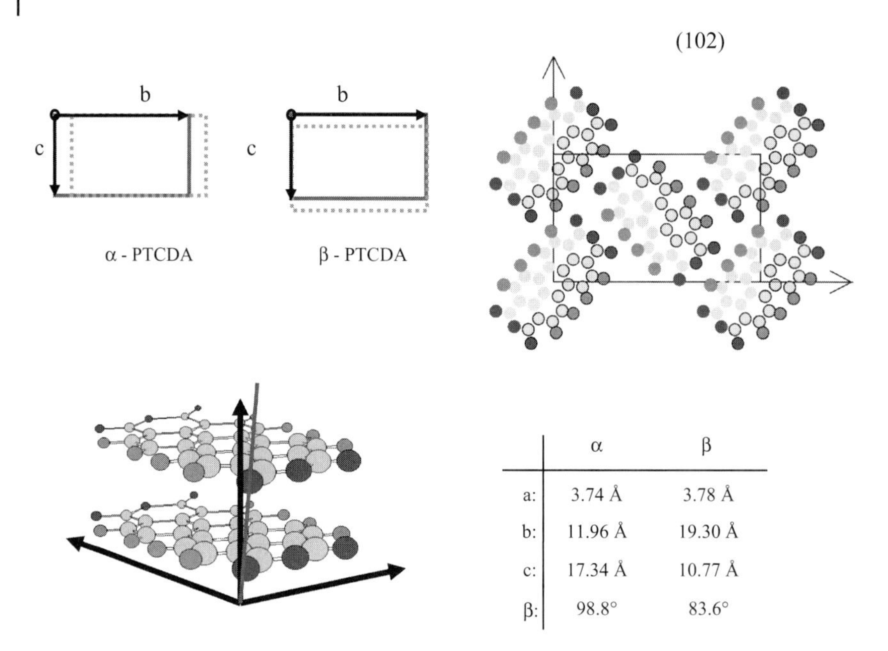

	α	β
a:	3.74 Å	3.78 Å
b:	11.96 Å	19.30 Å
c:	17.34 Å	10.77 Å
β:	98.8°	83.6°

Figure 3.1 Crystalline structures of PTCDA. Lattice parameters are from Reference [116].

observed from a single-domain-oriented homogenous monolayer with a "brick wall" structure into a more condensed "herring bone" structure [125]. The interfacial electron phonon-coupling, which may also be called interfacial dynamical charge transfer, is enhanced on Ag(111) surfaces by two orders of magnitude over a conventional chemisorptive bond as it is found for PTCDA on Ag(110) [126]. For Ni(111) substrates, PTCDA molecules have been found to adsorb with their molecular plane being tilted with respect to the substrate surface [127].

Compared to surfaces of metal single crystals, semiconductor surfaces seem to be more reactive substrates. As in the case of PTCDA grown on Ni(111), the molecules are found to be tilted on Si surfaces [127]. On Si(111)-(7 × 7) surfaces, a local interaction between the carboxylic groups of PTCDA and the dangling bonds on the substrate is found [128]. The π-bond within the C=O bond is lost. The PTCDA molecules adsorb in random orientations resulting in disordered layers [129]. A more ordered growth of PTCDA on Si is achieved by passivating the Si surface with hydrogen, which results in a reduced interaction between the molecule and the substrate surface. Here, the homogeneity of the substrate surface on a microscopic scale determines the structural quality of the organic films. For example, epitaxial growth of metal phthalocyanines on S(111) surfaces was only observed on surfaces exhibiting monohydrides, that is, on surfaces with a negligible density of steps [130].

On InAs(001) substrates, one monolayer of PTCDA interacts relatively strongly with the substrate and 2D overlayers are formed [131]. For larger coverages, a

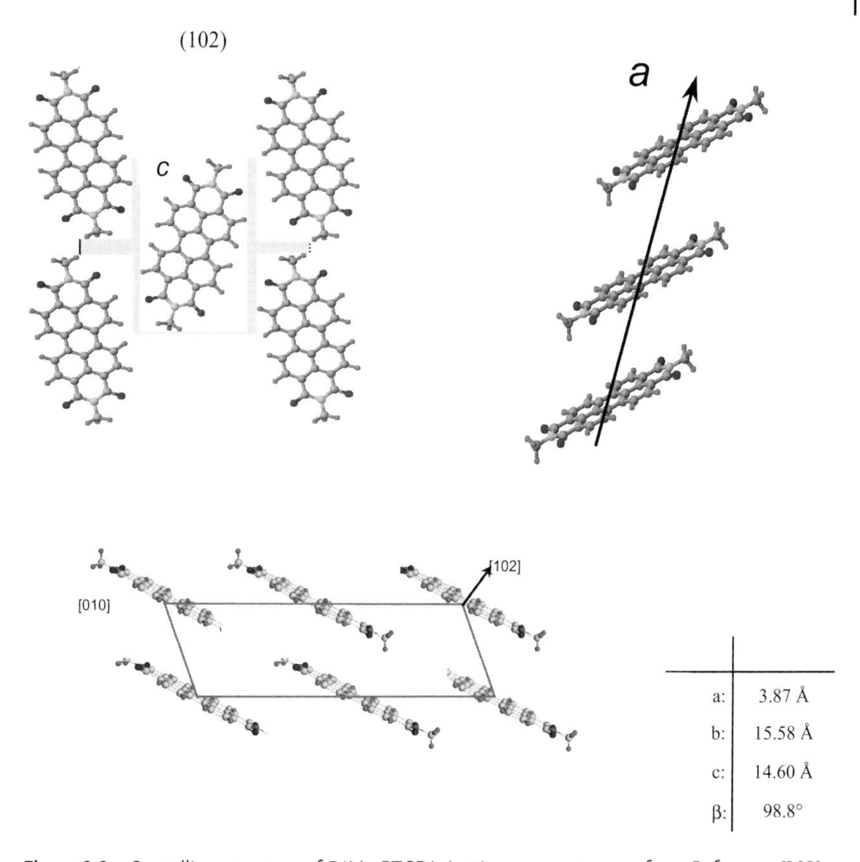

a:	3.87 Å
b:	15.58 Å
c:	14.60 Å
β:	98.8°

Figure 3.2 Crystalline structure of DiMe-PTCDI. Lattice parameters are from Reference [121].

phase transition occurs where bulk-like 3D PTCDA clusters begin to form. As for the case of elemental semiconductor surfaces, the strong interaction between PTCDA and compound semiconductor surfaces is considerably reduced by a passivation of the surface prior to the growth of the organic film [53, 54]. For III–V semiconductors, particularly GaAs(100) surfaces, various methods of surface treatment have been developed to improve the chemical and electronic properties of the surfaces. One of the most widely used methods for the passivation of the GaAs(100) surface is chalcogen passivation using S, Se, and Te atoms. Hirose *et al.* investigated the initial growth of PTCDA molecules on various GaAs(100) surfaces using low-energy electron diffraction (LEED) [53, 132]. The Se-passivated GaAs(100)-(2×1) surface results in PTCDA films with good crystallinity, whereas the deposition on the GaAs(100)-(2×4) or c(2×8) reconstructed surfaces produces films with crystallites randomly oriented in the plane parallel to the surface. The films grown on the GaAs(100)-c(4×4) surface show intermediate molecular ordering. The difference in the molecular ordering depends on the relative interaction between organic molecules to that between organic molecules and the substrate surfaces. In most of the examples described earlier, PTCDA films grow with the (102) plane being parallel to the substrate surface. Only if the substrate–molecule interaction is sufficiently strong, the orientation of the molecules will be changed. As for Ni and Si surfaces, this has been observed for the adsorption of PTCDA on alkali-halogenide substrates, where the anhydride groups of the molecule interact strongly with the substrate [133].

Changing the structure of the molecule, for example, by an imide group with different side groups instead of the anhydride group, the growth properties are changed drastically. Depending on the preparation procedure, di(2,6-isopropylphenyl)-3,4,9,10-perylenetetracarboxylic diimide molecules lie on the surface with the perylene core being parallel to or tilted with respect to the Ag(111) substrate [134]. In epitaxial films grown on cleaved KCl(001) surfaces, the molecular planes of DiMe-PTCDI molecules are co-facially stacked parallel to the KCl substrate, whereas dibutyl-3,4,9,10-perylenetetracarboxylic diimide molecules are oriented standing upright on the surface [135]. In thick films (300 nm) of DiMe-PTCDI grown on glass substrates, most of the molecules are parallel to the substrate, some of them being tilted with a Gaussian distribution having a FWHM of 22° [136]. A large tilt angle of about 90° was observed by means of LEED and scanning tunneling microscopy (STM) for monolayer films of DiMe-PTCDI on Ag(110) substrates [134] where a pronounced in-plane optical anisotropy was also detected [137]. Even on the same substrate, for example, Ag(110), various orientations can coexist [138].

3.1 Scanning Probe Microscopy

In scanning probe microscopy (SPM), a tip of preferably atomic sharpness scans across the surface. There are two techniques in SPM, which can be distinguished by the way the tip interacts with the surface.

Figure 3.3a shows the schematics of a typical STM. The tip is mounted on a piezoelectric tripod. With this piezoelectric tripod-sample, a sample–tip distance of the order of a few angstrom is established, where electrons may tunnel across the tip–sample gap upon applying a bias. Tunneling is a quantum mechanical effect where electrons penetrate a classically impenetrable potential barrier between two conductors. From a phenomenological point of view, this process can be described by the "leaking" of wavefunctions into the vacuum and thereby crossing the vacuum gap. This is only achieved for the atomically small distances. In a first classical approximation, the tunneling current I_T can be described by the following expression [139]:

$$I_T \propto \frac{U_T}{d}\exp\left(-Kd\sqrt{\varphi}\right) \tag{3.1}$$

where U_T is the voltage applied between tip and sample, ϕ their average workfunction, and K a constant with a value of $1.025\,\text{Å}^{-1}(\text{eV})^{-1/2}$ for a vacuum gap. The tunneling current I_T depends exponentially on the sample–tip distance.

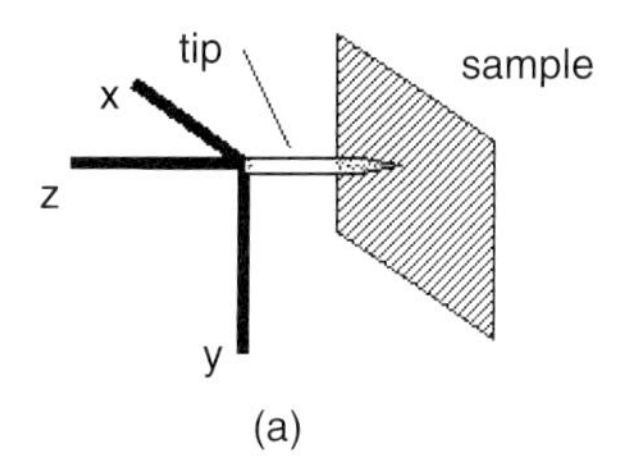

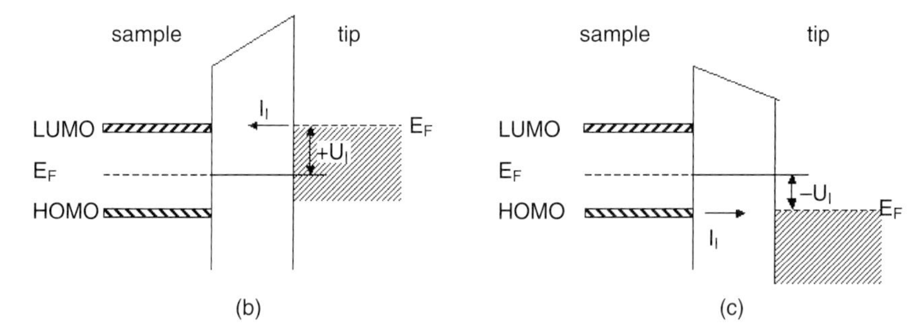

Figure 3.3 (a) Schematic of a typical STM. The tip is mounted on a piezoelectric tripod and scanned across the sample surface. (b, c) The electronic band diagrams for the sample and the tip under biases with different polarity. (b) The tunneling of electrons from the tip into the LUMO of the sample. In (c), the bias is reserved and electrons tunnel from the HOMO into the metal.

Tunneling occurs when wavefunction tails of occupied states on one side of the gap overlap with wavefunction tails of unoccupied states on the other side of the gap. This is achieved by varying the tunneling voltage U_T. In the metal tip, states close to the Fermi energy participate in the tunneling. Figure 3.3b shows the case where the Fermi level of the metal tip aligns with the LUMO of the sample. In this case, electrons tunnel from the tip to the sample. The tunneling current is reversed for a negative U_T where the Fermi level of tip aligns with HUMO of the sample (see Figure 3.3c). The tunneling occurs for $U_T < \phi$.

Once the tunneling is established, the tip is scanned in the *x*-, *y*-direction across the sample surface. Here, two operational modes are distinguished. On flat samples, the constant height mode can be used, where the sample–tip distance is kept constant and the tunneling current is recorded as a function of position. On surfaces with steps, the constant current mode is used. Here, the tunneling current is kept constant and the height is recorded as a function of position.

Another type of SPM technique is atomic force microscope (AFM), which images the force between tip and sample while scanning the surface. This technique does not rely on the electrical conductivity of the sample and can be used for insulators as well. There are two different AFM techniques. In contact AFM, a cantilever is excited at its resonance frequency and the amplitude of the oscillation is measured by deflecting a laser beam on the oscillating cantilever. The

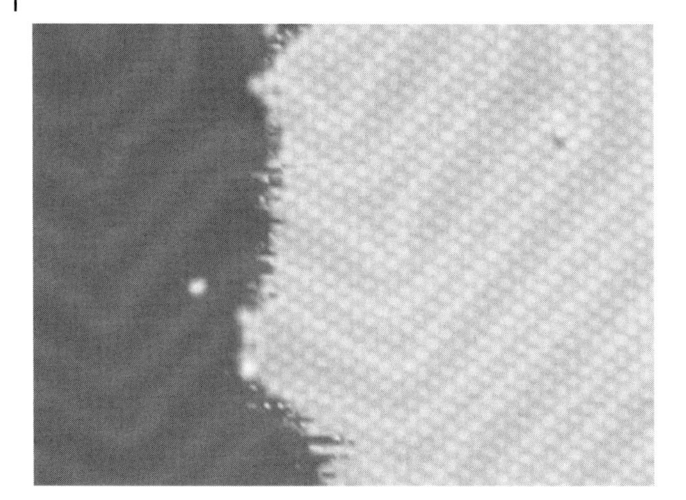

Figure 3.4 Constant current STM image of submonolayer PTCDA coverage on Au(111). (57 nm × 40 nm, $V = -2.0$ V, $I = 0.16$ nA).

change in the amplitude is a measure of the force between the cantilever tip and the sample. In this mode, the cantilever tip is permanently (contact mode) or periodically in contact with the sample. Since most of the molecules interact weekly with the substrate, this technique may move/drag molecules along the surface. In noncontact AFM, the vibrating tip is not in contact with the surface. Here, an electromechanical resonator is excited at its resonance frequency too, but the change in frequency is a measure of the force between tip and sample. To achieve the atomic resolution, these resonators are used in the so-called short-range interaction region, where the sample–tip distance is about 1 Å. Because of this small distance, amplitudes are well below 1 Å.

3.1.1 STM on PTCDA/Au(111)

Figure 3.4 shows the constant current STM image of PTCDA grown on Au(111) [140]. The coverage is below one monolayer and both the substrate surface and the thin organic film on top of it can be observed simultaneously. The left side, which appears darker, corresponds to the clean substrate. It shows the typical $22 \times \sqrt{3}$ surface reconstruction of the clean Au(111) surface.

The right side of the image corresponds to the thin layer of PTCDA, which is just one molecule "thick". The Au(111) surface reconstruction can still be observed through the PTCDA layer. The molecules in the PTCDA layer assemble in the herringbone pattern of the (102) plane of the PTCDA crystal. The edge of the PTCDA layer appears fuzzy, which is due to single molecules continuously absorbing and desorbing at the edge of the PTCDA layer while the image is taken. Single molecules cannot be imaged due to their high mobility on the surface. The fact that the substrate surface reconstruction is not influenced by the molecules, and that the molecules have a high surface mobility and assume their crystalline arrangement within the first monolayer indicates that the substrate–molecule interaction is very weak.

A further implication of this weak molecule–substrate interaction is shown in Figure 3.5. Here, STM images have been taken from a monolayer of PTCDA on Au(111) at RT. Figures 3.5a and c are taken from one and the same area showing the unoccupied states and occupied states with positive and negative sample bias, respectively, with the difference in sample bias being equivalent to the HOMO-LUMO gap in PTCDA. Both images show submolecular features and are com-

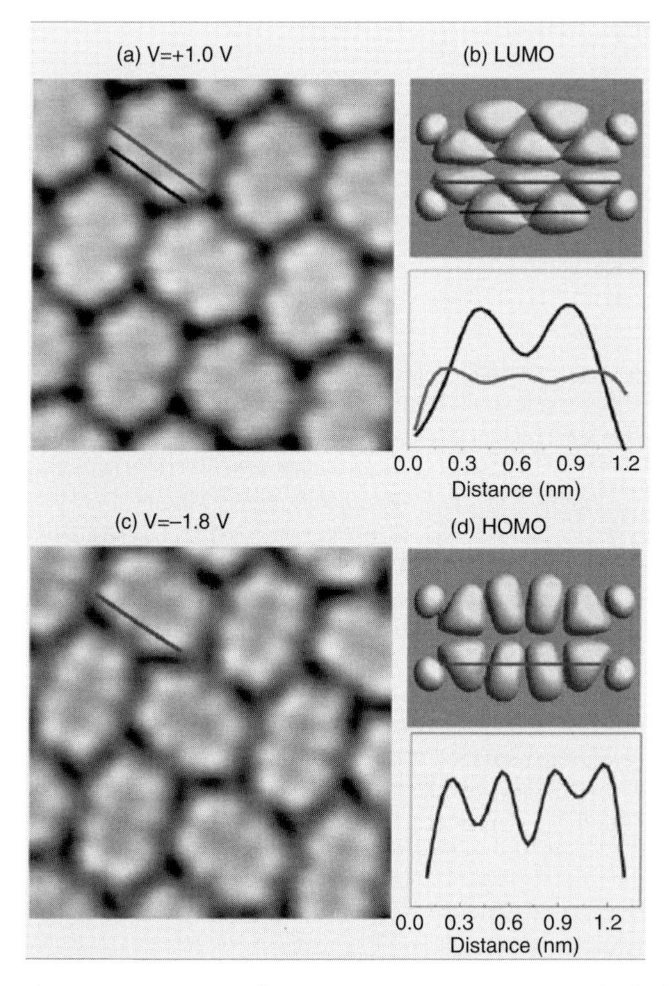

Figure 3.5 STM images of 1 ML PTCDA on Au(111) at RT obtained at a sample bias of +1.0 V (a) and $V = -1.8$ V (c). Image size and tunneling current are 4.2 nm × 4.2 nm and 0.6 nA, respectively. (b, d) Calculated HOMO and LUMO for the free molecule. Profiles along a molecule are shown for better comparison. [image from Reference 140]

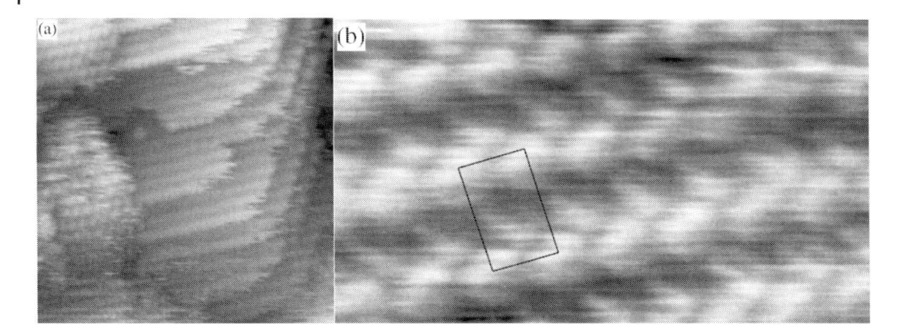

Figure 3.6 STM images ($I = 0.08$ nA, $V = -3.3$ V) after deposition of 14 ml PTCDA at 180 °C. (a) 44 nm × 44 nm; (b) 10 nm × 5.8 nm.

pared to the calculated HOMO and LUMO. These calculations were done within density functional theory (DFT) in the local density approximation (LDA) [141], using the SIESTA method [142, 143]. There is an excellent agreement between the experimental and theoretical data. For a more detailed comparison, experimental line profiles are compared with the number of maxima to be expected from the calculated images.

3.1.2
STM on PTCDA/S-passivated GaA(100)

An example of PTCDA films, which are even highly ordered along the growth direction, is shown in Figure 3.6. The STM images show 14 ML of PTCDA grown on sulfur-passivated GaAs(001) [144]. The sulfur passivation drastically decreases the chemical reactivity of the GaAs surface so that this surface may be considered passivated, that is, chemically inert. Due to this passivation, the PTCDA–GaAs(001) interaction is drastically reduced and the molecules show a highly ordered 3D growth. To support this highly ordered growth, the substrate temperature had been at 180 °C during the deposition of the PTCDA. Figure 3.6a shows the terraces of a PTCDA crystal where the herringbone arrangement is observed on each terrace. This is more easily seen on the more detailed image in Figure 3.6b, which has been taken from one of the terraces. It can be concluded from these images that the (102) plane of the PTCDA crystallite is parallel to the (001) surface of the GaAs substrate. The internal ordering of these films is high enough to provide enough conductivity for taking the molecular resolved STM images.

3.1.3
AFM on PTCDA and DiMe-PTCDI/S-passivated GaAs(100)

Figure 3.7 shows a topographic AFM picture (1 μm × 1 μm) taken from a 30-nm thick PTCDA film. The AFM [145] images were obtained with a commercial microscope (Nanotec Electronica S.L., Spain) equipped with Olympus type

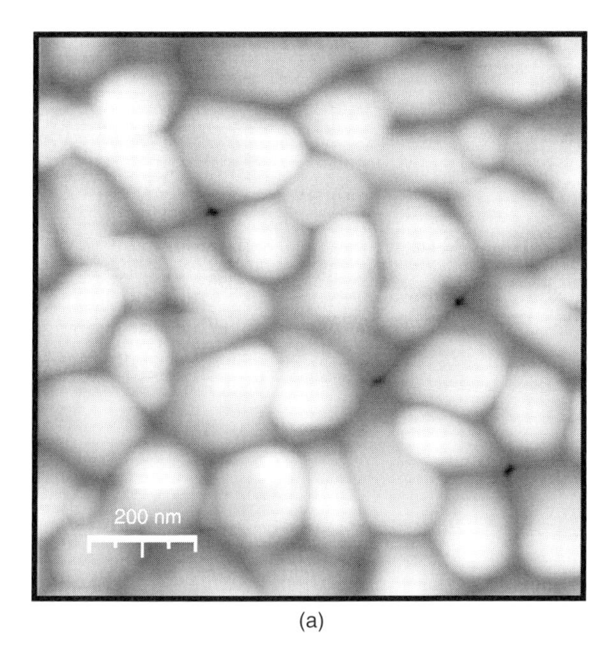

(a)

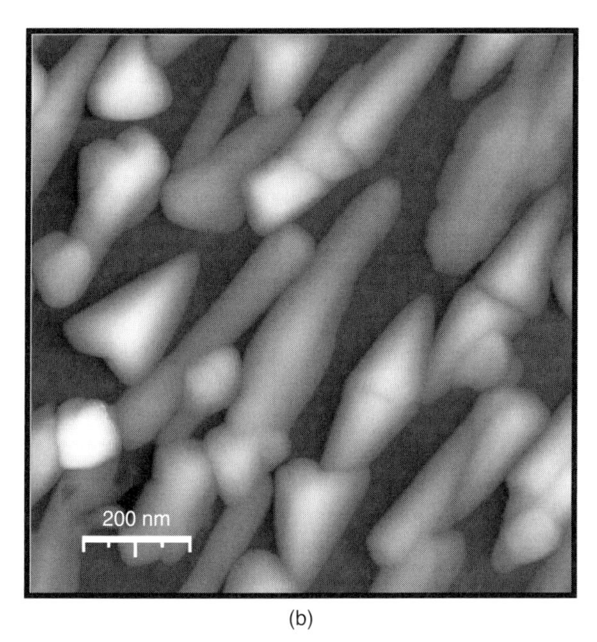

(b)

Figure 3.7 AFM images of (a) PTCDA and (b) DiMe-PTCDI films grown on S-passivated GaAs(100). [292]

cantilevers having a 15-nm nominal radius, a spring constant of 1 N/m, and a resonance frequency of 78 kHz. AFM images were recorded in noncontact tapping mode using the conditions described in detail elsewhere [146]. Typically, the microscope was operated in ambient conditions at a relative humidity of 30%. Deposition of PTCDA on S-passivated GaAs(100) at room temperature leads to PTCDA aggregates with dimensions between 50 and 200 nm. In Figure 3.4a, topographic AFM picture (1 μm × 1 μm) corresponding to a nominal coverage of 20 nm is shown. In the case of DiMe-PTCDI on S-passivated GaAs(100), deposition at room temperature leads to the formation of ribbon-like crystals with a width of about 50–100 nm and a length of about 100–600 nm. The crystals in the image are aligned with their long axis being parallel to one of the edges of the cleaved substrate.

3.2 X-Ray Diffraction (XRD)

X-ray diffraction is applied to bulk material, thin films, and powders to yield information on crystalline structure and orientation, crystalline perfection, composition, thin film thickness and uniformity, as well as strain. The Bragg formulation of X-ray diffraction is based on Bragg's law [147]:

$$2d_{hkl}\sin\theta_B = n\lambda, \text{ with n} = 1, 2, 3, \ldots \tag{3.2}$$

Electromagnetic waves with a wavelength comparable to interatomic distances are scattered at lattice planes with Miller indices (hkl) and spacing d_{hkl}. Constructive interference occurs where $2d_{hkl}\sin\theta_B$ is equal to multiples of the wavelength, or, in other words, when the phase shift is a multiple of 2π. Here, θ_B is the respective scattering or Bragg angle.

Figure 3.8a shows the scattering geometry for a crystal where the lattice planes (hkl) of a crystal are tilted by an angle φ with respect to the surface. The incident and scattered wave vector, and the normal to the lattice plane are denoted by $\mathbf{k}_i$, $\mathbf{k}_s$, and $\mathbf{n}$, respectively. The scattering angle θ is the angle between $\mathbf{k}_i$ and the lattice planes. It should be noted that for clarity Figure 3.8 shows the scattering geometry at the surface, while in reality X-rays have a large penetration depth and scattering takes predominantly place in sub-surface regions.

For the general situation shown in Figure 3.8a, where $\varphi \neq 0$ results in $\omega_+ \neq \omega$, the Bragg diffraction is called "asymmetric" and is defined by a asymmetry factor

$$b = \frac{\sin(\omega_+)}{\sin(\omega_-)} = -\frac{\sin(\theta+\varphi)}{\sin(\theta-\varphi)} \tag{3.3}$$

where ω_+ and ω_- are the angles between the surface plane and the incident and scattered wavevector $\mathbf{k}_i$ and $\mathbf{k}_s$, respectively. For the reflecting lattice planes being parallel to the surface $\varphi = 0$ the Bragg diffraction is called "symmetric". This situation, where $\omega_+ = \omega_- = \omega = \theta$, is illustrated in Figure 3.8b. To do an angular scan the sample is rotated, but the detector has to be rotated twice as fast in the same direction.

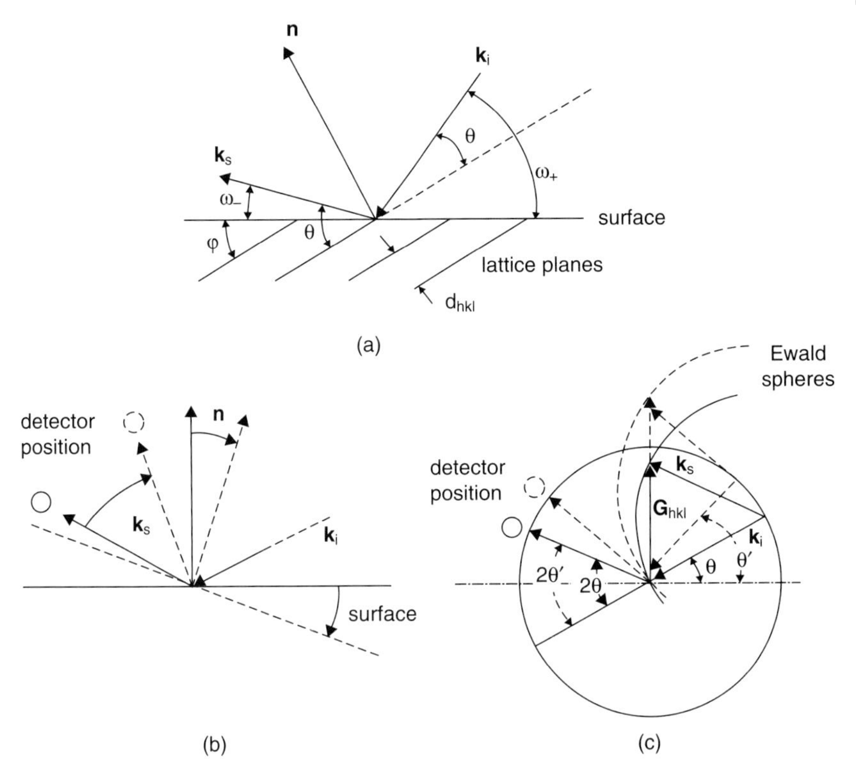

Figure 3.8 a) Scattering geometry in x-ray diffraction. b)Geometry for symmetric Bragg scattering. c) Ewald sphere construction for the θ-2θ-scan geometry. Dashed lines illustrate the geometry after rotating the sample.

In reciprocal space a set of lattice planes is described by a reciprocal lattice vector $\mathbf{G}_{\mathrm{hkl}} = 2\pi/2d_{hkl}$ which is normal to the set of planes. Then Bragg's law expresses the momentum conservation:

$$\mathbf{k}_s = \mathbf{k}_\mathrm{i} + \mathbf{G}_{\mathrm{hkl}} \tag{3.4}$$

The conditions for Bragg diffraction in a $\theta - 2\theta$ scan geometry can be determined from the corresponding Ewald sphere construction shown in Figure 3.8c. For elastic scattering $\mathbf{k}_\mathrm{i}$ and $\mathbf{k}_\mathrm{s}$ have equal length and their tips lie on the surface of a sphere with $2\pi/\lambda$ radius. Diffraction occurs when the origin of $\mathbf{G}_{\mathrm{hkl}}$ is placed on the tip of $\mathbf{k}_\mathrm{i}$ and it's tip falls on to the surface of the Ewald sphere. During a $\theta - 2\theta$ scan the tip of $\mathbf{k}_\mathrm{s}$ moves along the reciprocal lattice vector $\mathbf{G}_{\mathrm{hkl}}$.

3.2.1 XRD on PTCDA grown on GaAs and Si

In the following, XRD data taken from PTCDA grown on sulfur-passivated GaAs(110) are discussed. The measurements have been performed using a SEIFERT 3000 PTS diffractometer with Cu Kα($\lambda = 1.5415$ Å) radiation. The X-ray generator is operated at 40 kV and 20 mA. A bending graphite monochromator is placed at the detector side. Soller slits were also included at both sides of the incident and reflected beams to limit any axial divergence. The reflected beam was detected through a 0.1-mm receiving slit for reflectivity studies. For recording the diffraction spectrum, θ–2θ scans in the range of 5–80° with a step width of 0.02° were carried out in Bragg–Brentano geometry. The SEIFERT X-ray diffraction software containing a reflectivity simulation program is used for the simulation of X-ray reflectivity curves.

Figure 3.9 shows the XRD spectra of films deposited on H–Si and S–GaAs substrates. The sublimed PTCDA powder shows many diffracting Bragg planes and those planes can be indexed following the cell parameters published in literature [148]. However, there is only one recordable diffracted peak corresponding to the (102) plane for PTCDA thin films deposited on the H–Si and S–GaAs surfaces at $2\theta \sim 27.65°$. The full width at half maxima (FWHM) of the (102) diffracting plane is found as ~0.34° for PTCDA films on both substrates when fitted with a single Gaussian peak. The FWHM of the same XRD peak of α-PTCDA single crystal is ~0.15° [149]. When the Scherrer's formula [150] is adopted for the evaluation of grain size in the films, a value of ~45 nm is obtained. In the present case, the FWHM of the single crystal PTCDA is taken as the standard peak broadening for the determination of grain size.

From AFM images, the particle size is of the order of 80–100 nm. The imparity in size could be due to a possible overlapping among the domains in the AFM image. However, one finds the coexistence of both phases in PTCDA films by Raman spectroscopy as discussed in Section 4.2. Moreover, a prominent asymmetry for the spectrum grown on hydrogen-passivated Si(100) toward lower 2θ value is clearly seen in Figure 3.9. In addition, a relative shift in the peak position by ~0.08° occurs for the XRD spectrum of PTCDA film on the H–Si surfaces in comparison to the film on S–GaAs surfaces. The monoclinic PTCDA(102) plane for α- and β-phases appears at $2\theta = 27.81$ and 27.44°, respectively, in accordance

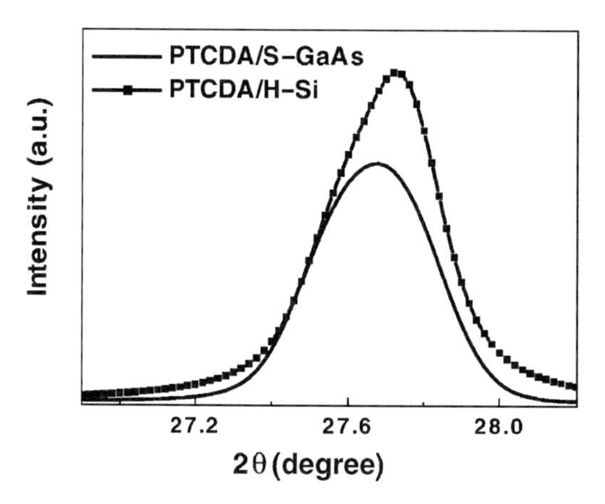

Figure 3.9 The (102) diffraction peaks of PTCDA films grown at room temperature.

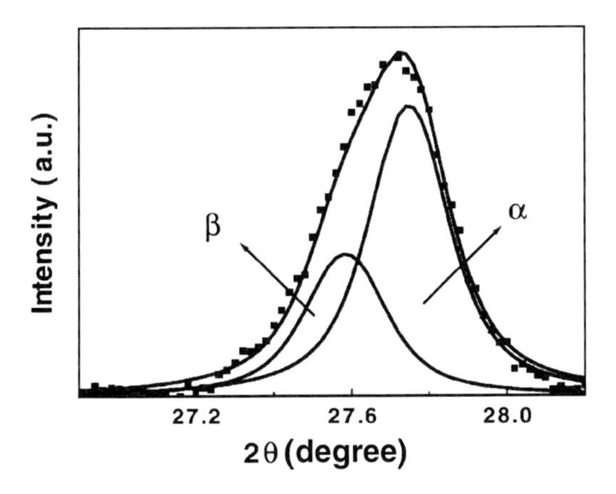

Figure 3.10 Voigt fit for the (102) diffraction peak of PTCDA film deposited on H–Si surfaces.

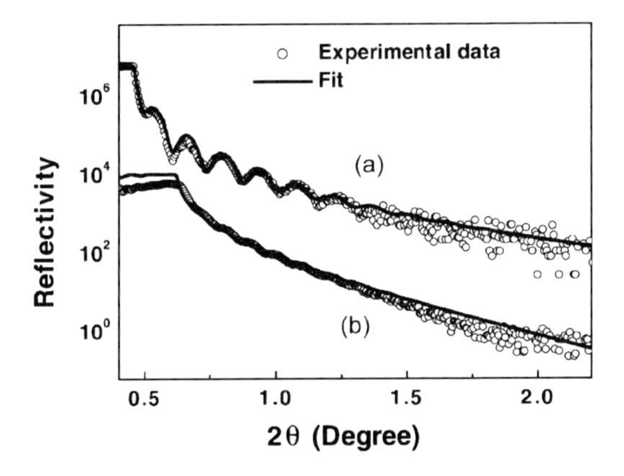

Figure 3.11 X-ray reflectivity of PTCDA films deposited on (a) H–Si and (b) S–GaAs substrates.

to the cell parameters reported for single crystal PTCDA [148]. Noteworthy, the XRD peak positions for thin films may depend on the interface properties as well as on the growth conditions, which can impart strain in the film [151]. As a result, this can shift the peak position in thin films. Taking into account the possible presence of both phases, a curve fitting is employed to get further information. Figure 3.10 shows the Voigt fitting of a typical XRD spectrum of a PTCDA film deposited on H–Si substrate into two peaks centered at 2θ values 27.58° and 27.78° having FWHMs of ~0.25°. The corresponding grain size is now ~80 nm, which is very close to the observable domains in the AFM image. Similarly, the curve fitting for the PTCDA film grown on an S–GaAs substrate reveals two peaks centered at 27.55° and 27.74° and the calculated grain size is ~75 nm. Considering the peaks at lower and higher 2θ values, which originate from β- and α -phases of PTCDA, respectively, the relative intensity ratio of α- and β-phases are determined to be 1.56 and 2.06 for PTCDA films on S–GaAs and H–Si substrates, respectively.

Figure 3.11 shows the reflectivity curves measured for the PTCDA/H–Si and PTCDA/S–GaAs systems. For X-rays, the index of refraction n is conventionally taken as $n = 1 - \delta + i\beta$, where δ is proportional to the electron density of the material and β to the absorption. A typical experimental reflectivity curve is calculated with the Parrat's formalism for two interfaces, taking into account the roughness [152]. The best parameters obtained by a fitting procedure are $d = 56.5 \pm 0.5$ nm for the thickness of PTCDA films on H–Si substrate and $\delta_S = (7.8 \pm 0.1) \times 10^6$ for the refractive index of the silicon substrate. The latter was set to the bulk value of 7.44×10^6 for the deviation of other parameters such as density and absorption coefficient [153]. For the PTCDA film on H–Si surfaces, $\delta_{PTCDA} = (4.1 \pm 0.1) \times 10^6$ is obtained from the simulation of the reflectivity curves. $\delta_{PTCDA} = (4.5 \pm 0.1) \times 10^6$ has been reported for films deposited on a silicon substrate in the literature [154]. For a film deposited on S–GaAs substrate, $d = 81.5 \pm 0.5$ nm and $\delta_{PTCDA} = (3.5 \pm 0.1) \times 10^6$ are determined. For the GaAs substrate, δ_S is found to be $(15.1 \pm 0.1) \times 10^6$ which agrees well with the reported value of 15.1×10^6 [153]. It should be mentioned that the theoretical bulk δ_{PTCDA} value is 5.6×10^6. In the earlier report on overall smaller δ values for films on Si substrates and consequently a smaller density than bulk, PTCDA is being argued as an outcome of the presence of amorphous PTCDA [155]. The roughness of the PTCDA–air interfaces is found to be 1.9 ± 0 and 2.2 ± 0.1 nm for films deposited on H–Si and S–GaAs, respectively. The interface roughness of the PTCDA/H–Si and PTCDA/S–GaAs systems are found to be 0.4 ± 0.05 and 0.7 ± 0.05 nm, respectively. The measured thickness and surface roughness of the PTCDA films on both substrates using AFM closely match with the X-ray reflectivity results. Möbus and Karl [155] reported that δ_{PTCDA} for films on Si substrates was always determined to be $(4.5 \pm 0.1) \times 10^6$ and there was no discernible trend with the deposition temperature. In the present case, the δ_{PTCDA} value obtained for films on H–Si surfaces $(4.1 \pm 0.1) \times 10^6$ is found to be slightly higher than that obtained for films deposited on S–GaAs surfaces $(3.5 \pm 0.1) \times 10^6$. As a consequence, a relatively higher density results for the PTCDA films grown on H–Si. A higher density for PTCDA films on passivated Si(111) substrates compared to films on GaAs substrates has been reported using optical spectroscopy [156]. This observation can be explained as follows. Both crystalline phases of PTCDA have two molecules per unit cell. However, considering the crystalline monoclinic cell parameters, the α-phase has a lower unit cell volume (762 Å^3) than the β-phase (780 Å^3) [148]. This clearly indicates that the α-phase is comparatively denser than the β-phase. As a result, the ratio of α- and β-phases can, to some extent, influence the overall density of the film. As we observe the preferential growth of the a-form on the H–Si substrates from the XRD measurements as well Raman spectral analysis, it is thus obvious that PTCDA films on H–Si surfaces should have a higher density. On the other hand, films on S–GaAs surfaces should show a lower density due to the higher contribution from b-PTCDA phase. Hence, due to the two different substrate surfaces, a distinct change in the growth mode of PTCDA phases occurs and importantly even by using low-intensity laboratory X-ray source facility, XRR measurements can bring out this influence of the substrate surfaces on the organic films.

Both types of organic films were investigated with XRD. Diffraction peaks could only be recorded for the PTCDA films. The XRD spectra for PTCDA grown S-passivated GaAs(100) presented in Figure 3.9 shows a Bragg diffraction peak around $2\theta \sim 27.65°$ corresponding to (102) planes of PTCDA and thus proves the crystalline nature of the islands observed by AFM. Moreover, this indicates that the (102) crystalline planes are close to parallel to the substrate plane. Recalling the crystal unit cell of PTCDA shown in Figure 3.1, such an orientation is equivalent with an orientation of the molecular planes close to parallel to the GaAs(100) surface.

The FWHM of the (102) diffraction peak w_f is found to be ~0.34° when fitted with a single Gaussian peak. The FWHM of the same XRD peak of an α-PTCDA single crystal w_c is 0.15° [149]. Mechanisms leading to a larger FWHM in films compared to a crystal are the finite size of crystalline domains and the existence of different polymorphs. The Bragg diffraction peaks of the monoclinic PTCDA (102) plane for α and β phases should appear at $2\theta = 27.81°$ and 27.44°, respectively, in accordance with the unit cell parameters reported for single crystal PTCDA. Noteworthy, the XRD peak positions for thin films may depend on the interface properties as well as on the growth conditions, which can induce strain in the film. As a result, the peak position in thin films can be shifted. Introducing $\Delta w = w_f - w_c$ in the Scherrer's formula, the average size of the crystalline grains ϕ can be estimated:

$$\phi = \frac{0.9 \times \lambda}{\cos(\theta) \times \Delta w} \tag{3.5}$$

with $\lambda = 1.5415\,\text{Å}$ being the wavelength of X-ray radiation.

Fitting for the XRD spectra with two Voigt profiles, an FWHM of ~0.3° is obtained. Using relation (5.1) gives a grain size of ~ 65 nm, which is in good agreement with the results from the AFM investigations. With the peak at low and high 2θ values assigned to the β- and α-phases of PTCDA, respectively, the relative ratio of α- and β-phases is determined to be 1.1 for PTCDA films on S-passivated GaAs(100). Using X-ray reflectivity, Das *et al.* determined the roughness of the PTCDA–air and the PTCDA/S–GaAs(100) to be 2.6 ± 0.1 and 0.7 ± 0.05 nm, respectively. The latter value is comparable to the roughness of 0.6 nm of similarly treated S-passivated GaAs(100) substrates determined by means of STM [157].

3.3 Low-Energy Electron Diffraction

In LEED, electrons with an energy of around 10–150 eV are reflected at a surface. Because of the low kinetic energy, this technique is very surface sensitive, and only the periodicity in the arrangement of the surface atoms is investigated. The periodic surface structure results in a diffraction pattern created by the backscattered electrons. This diffraction pattern is visualized on a phosphorous screen, which is behind an arrangement of hemispherical grids serving as a high-pass filter. Only elastically scattered electrons have sufficient kinetic energy to overcome the high-pass filter.

3.3.1
LEED on PTCDA/Au(111)

Figure 3.12 shows an LEED image taken from 0.5 ml of PTCDA grown on Au(111) [158]. The PTCDA was deposited with a deposition rate of 0.5 ml/min. The diffraction pattern can be understood as a superposition of six symmetry-equivalent domains of a rectangular adsorbate structure. The existence of six domains is a result of the number of possible orientations provided by the substrate surface

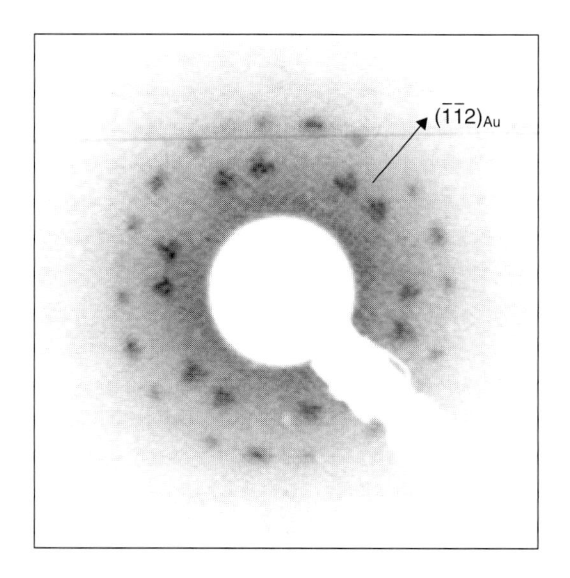

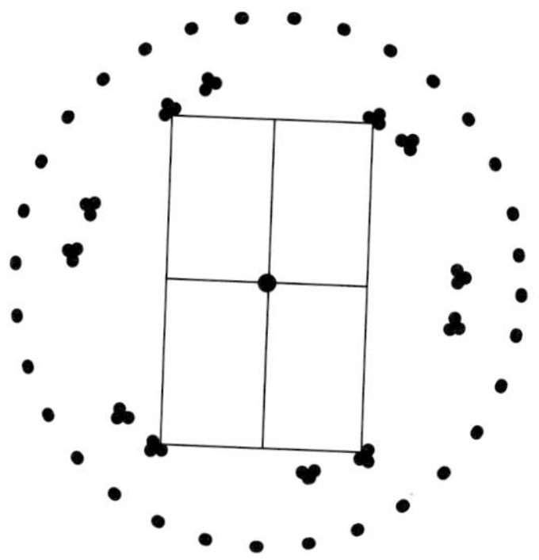

Figure 3.12 LEED image of 0.5 ml of PTCDA grown on Au(111). The kinetic energy of the electrons is 20.6 eV. Lower part shows a calculated diffraction pattern.

times the number of possible orientations of an adsorbate structure, which does not align with the substrate lattice: the $22\times\sqrt{3}$ reconstructed Au(111) surface has three domains and there are two symmetry equivalent mirror domains for the PTCDA. The calculated LEED pattern based on such a structure model is shown in the lower part of Figure 3.12.

Increasing the deposition rate further, LEED spots can be observed, which are due to a second phase with again six symmetry-equivalent domains. This phase has different lattice constants and covers a smaller fraction of the substrate surface. It disappears after annealing the sample at 470 K.

3.3.2
micro-LEED on Pentacene/Si(111)

The investigation with a standard LEED optics shows that due to the limited lateral resolution, the diffraction patterns are usually a superposition of the existing domains. A low-energy electron microscope (LEEM), on the other hand, achieves the necessary lateral resolution to investigate single domains in organic thin films with a size of below 100 nm. In this technique, electrons of about 15 keV kinetic energy are focused onto the sample. Applying a retarding bias to the sample decelerates the electrons, that is, the kinetic energy of the impinging electrons is set by the sample bias to typical energies used in a standard LEED. The diffracted electrons are accelerated by the sample bias and collected by an immersion lens with a large acceptance angle. The instrument can run in lateral as well as angular resolving lens modes, where the latter one can be used for so-called nano-LEED investigations.

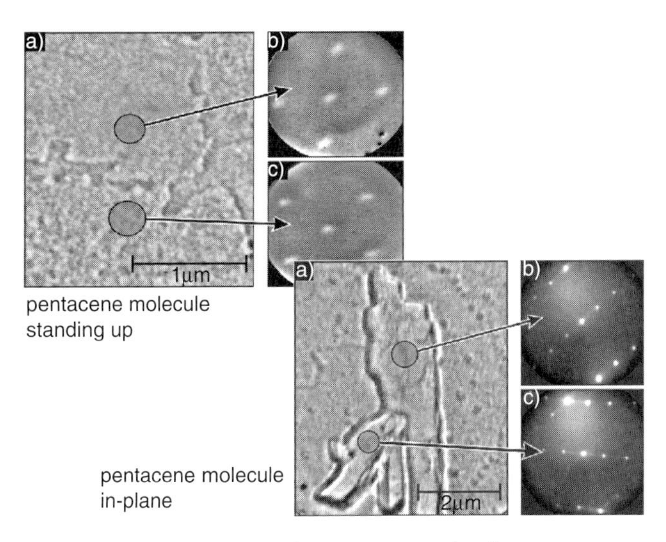

Figure 3.13 LEEM images and LEED patterns taken from pentacene grown Si(111). LEED patterns are taken from areas indicated by the circles.

Figure 3.13 shows images as well as nano-LEED patterns taken from pentacene grown on Si(111) [293, 294]. The pentacene islands have dimensions of a couple of micrometers. The lateral resolution of the LEEM is sufficient to investigate single domains. The LEED patterns taken from the two adjacent domains shown in the upper image indicate that here the pentacene molecules are standing upright. In the lower image, again two adjacent domains are shown. Here, the LEED patterns clearly indicate that the molecules are lying flat.

4
Optical Spectroscopy

In inorganic materials, next-neighbor interaction is strong due to chemical bonding resulting in the delocalization of the outermost occupied states. The evidence is the formation of valence and conduction bands whose energy width are several electron volts. In contrary, the weak intermolecular interaction in organic thin films and organic crystals leads to a highest occupied molecular orbital (HOMO) and a lowest unoccupied molecular orbital (LUMO), which are mostly localized at the respective molecule and show a bandwidth of around 0.1 eV. Therefore, the optical properties are in a first-order approximation similar to those of a single molecule, that is, a molecule in gas phase or solution, where a red shift and broadening of the optical spectra in the later case.

Absorption and recombination in a single molecule can be understood in terms of a displaced harmonic oscillator. In a classical picture, the potentials of the ground and excited state have a parabolic shape. Upon optical excitation, the excited molecule will change its geometry, which is reflected by a shift of the minimum of the excited state potential in space. The right-hand side of Figure 4.1 shows that the absorption will start by the excitation of an electron from the lowest vibrancy level of the ground state into one of the higher vibronic levels of the excited state. Vertical arrows indicate these transitions, since the configuration of the molecule is not changed at the instant of the transition. The molecule returns to the electronic ground state. The excess energy is converted to vibrational energy (*internal conversion*), and so the molecule is placed in an extremely high vibrational level of the electronic ground state. This excess vibrational energy is lost by collision with other molecules (*vibrational relaxation*). The spin of an excited electron can be reversed, leaving the molecule in an excited *triplet* state; this is called *intersystem crossing*. The triplet state is of a lower electronic energy than the excited singlet state. The probability of this happening is increased if the vibrational levels of these two states overlap. For example, the lowest singlet vibrational level can overlap one of the higher vibrational levels of the triplet state. A molecule in a high vibrational level of the excited triplet state can lose energy in collision with solvent molecules, leaving it at the lowest vibrational level of the triplet state. It can then undergo a second intersystem crossing to a high vibrational level of the electronic ground state. Finally, the molecule returns to the lowest vibrational level of the electronic ground state by vibrational relaxation. This is schematically shown in Figure 4.2.

Low Molecular Weight Organic Semiconductors. Thorsten U. Kampen

ISBN: 978-3-527-40653-1

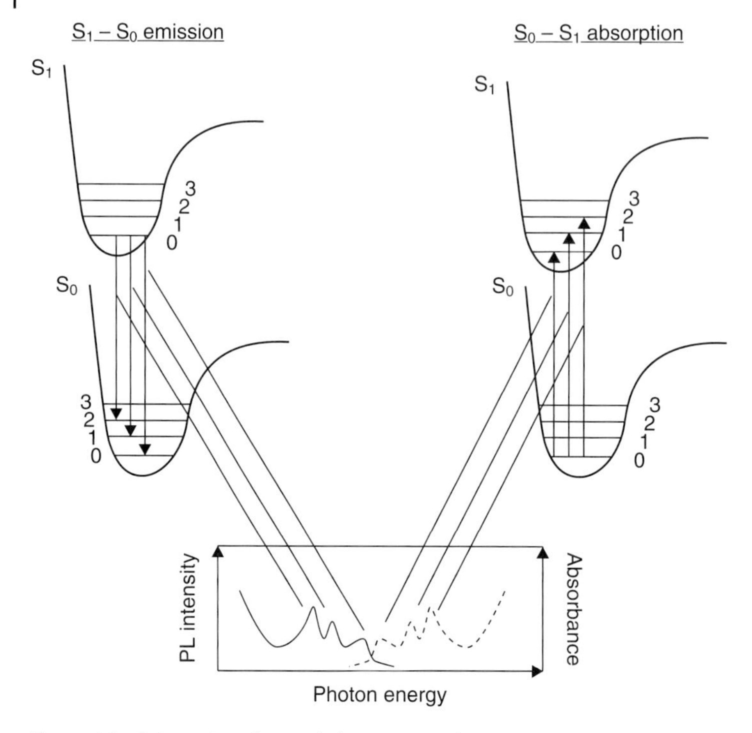

Figure 4.1 Schematics of optical absorption and emission.

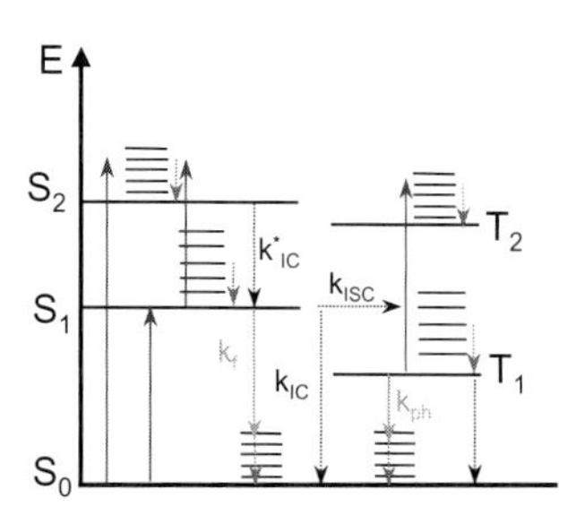

Figure 4.2 Energy level scheme of an organic molecule. S and T stand for singlet and triplet states, respectively. k_{IC}, k_{ISC}, k_f, and k_{ph} indicate transitions by internal conversion, intersystem crossing, fluorescence, and phosphorescence, respectively.

A molecule in the excited triplet state may not always use intersystem crossing to return to the ground state. It could lose energy by emission of a photon. A triplet/singlet transition is much less probable than a singlet/singlet transition. The lifetime of the excited triplet state can be up to 10 s, in comparison with 10^{-5} to 10^{-8} s average lifetime of an excited singlet state. Emission from triplet/singlet transitions can continue after initial irradiation. Internal conversion and other radiationless transfers of energy compete with phosphorescence.

Another implication of the strong localization of occupied and unoccupied states is that optical excitations are mostly localized on one molecule. The localized electron–hole pair is called an exciton and can be treated as a single particle. The concept of excitons was first introduced by Frenkel [159] and the term Frenkel excitons (FEs) is used for excitons with a small radius of about ~5 Å located at one molecule. In inorganic materials with delocalized states and efficient screening, excitons have a radius of several 10 Å and are called Wannier excitons [160]. The binding energies amount to about 10 meV.

In addition, self-trapped excitons like charge transfer excitons or excimer excitons can occur. These excitons are accompanied by charge transfer between next-neighbor molecules. If the time for the charge transfer is longer than the respective time for lattice deformation, the exciton is trapped in this deformation. The result is an anion–cation complex. An excimer (excited dimmer) exciton is the result of a dimmer formation between an excited molecule and a neutral molecule.

The asymmetry in the structure of molecular crystals and thin films results in anisotropic physical properties. Zang *et al.* [161] and Friedrich *et al.* [162] demonstrated the anisotropic optical properties of thin PTCDA films and the pronounced differences for in-plane and out-of-plane optical constants. Hädicke *et al.* [163] reported the solid-state effect of crystallochromy for *N,N′*-substituted-3,4,9,10-perylenetetracarboxylic diimide (R-PTCDI) with different peripheral R-groups like methyl or butyl. Crystallochromy is the change in absorption line-shape of films (crystals) compared to the monomer in solution and has been observed for a large number of perylene derivatives [164]. This change can be related to the transfer of FEs between neighboring molecules [165], considering a regular arrangement of the molecules in the crystalline domains having finite size. The absorption spectra corresponding to transitions between the HOMO and the LUMO of R-PTCDI in the crystalline state shift depending on the electronic interaction between the close-packed perylene planes. Different steric effects of the substituted R-groups can vary the interaction.

The strong interaction between next-neighbor atoms in inorganic semiconductors results in the development of energy bands. The electrons within this energy bands are delocalized and current transport can be described by free electrons. Carbon chains and rings determine the molecular and electronic structure of organic semiconductors. Here, the outer valence electrons occupy 2s and 2p orbitals. For the formation of benzene rings, sp^2 hybrid orbitals are formed resulting in three orbitals within a common plane and having an angle of 120° with respect to each other. Due to the overlap of these orbitals with the valence orbitals of next-neighbor atoms, localized σ-bonds between the atoms are formed. The remaining p_z orbital, which does not participate in the formation of the sp_2 hybrid orbitals, is oriented perpendicular to the plane consisting out of the atoms and the σ-bonds. The overlap of these p_z orbitals results in the π-bonds in which the electrons are delocalized above and below the molecular plane. The electrons in the π-orbitals occupy the HOMOs and are, therefore, important for charge transport and electronic excitation processes. The antibonding π-orbitals are denoted as π^*-orbitals. The π–π^*-transition between the HOMO and the LUMO is the transition with the lowest excitation energy. For example, the HOMO and LUMO of PTCDA and DiMe-PTCDI calculated using a density functional (DF) method of Gaussian'98 (B3LYP, 6-31G(d)) [166] are shown in Figure 4.3.

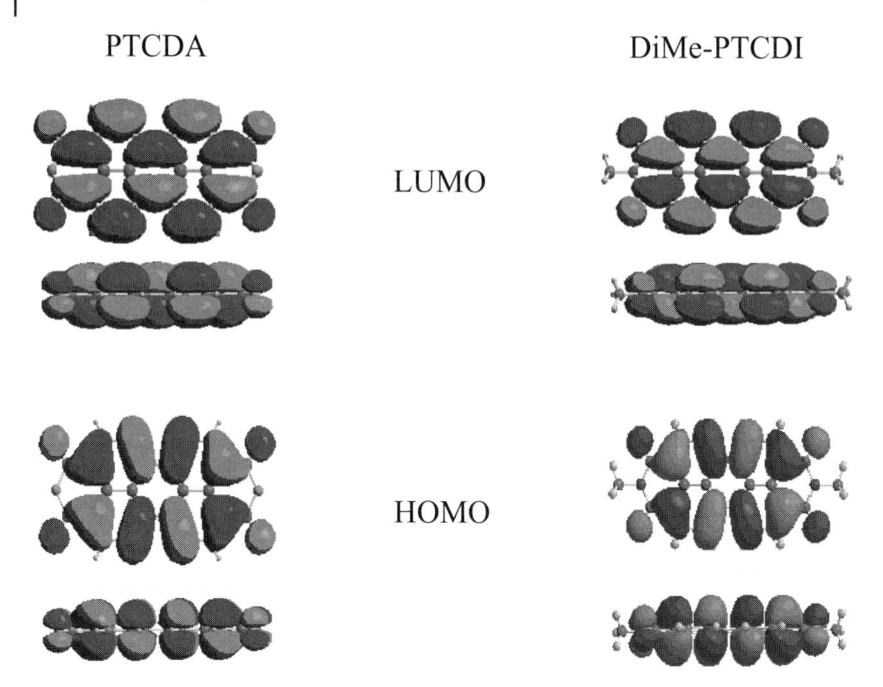

Figure 4.3 Top and side view of HOMO and LUMO of PTCDA and DiMe-PTCDI.

Due to the filled π-electron system, next-neighbor molecules interact via van der Waals forces. The charge carriers are localized at the respective molecules. The electronic and optical properties are, therefore, predominantly determined by the electronic properties of the molecule.

The optical absorption coefficients of PTCDA and DiMe-PTCDI thin films grown on quartz substrates are shown in Figure 4.4. The optical absorption is due to dipole-allowed transitions between occupied and unoccupied states where S_0 is for HOMO, S_1 for LUMO, and S_2 for the next excited state accessible by a dipole-allowed transition. The transition at the lowest energy is E_{0-0}, which is a transition between the ground vibrational states of two electronic levels. For PTCDA and DiMe-PTCDI molecules in solution and thin films, the vibronic progression of the S_0–S_1 transition with the E_{0-0} peak dominates the spectra. In thin films grown on quartz substrates, the S_0–S_1 transitions are red-shifted compared to the monomers in solution and the first absorption peaks appear at 2.22 and 2.14 eV for PTCDA and DiMe-PTCDI, respectively [167, 168] (Section 4.3.1). These energy values are the optical band gaps. Moreover, the broadening is so large that the individual vibronic peaks are not resolved.

Strong polarization effects characterize the electronic and electrical properties of an organic solid [169]. These polarization effects result in a shift of the energy levels in a molecular solid with respect to their energy position in single molecules in the gas phase. The principal polarization effects to be considered when describ-

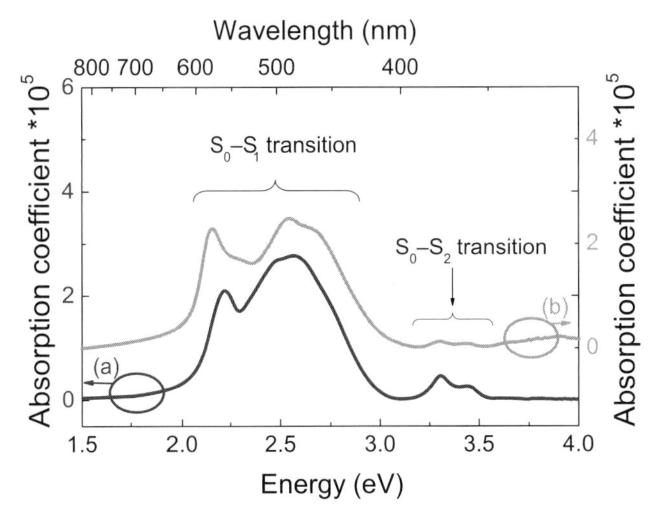

Figure 4.4 Absorption spectra of PTCDA (a) and DiMe-PTCDI (b) films (50 nm each) deposited on quartz substrates at room temperature (with the courtesy of A. Paraian).

ing the interaction of an ion with its next-neighbor molecules are electronic polarization, intramolecular polarization, and lattice polarization [170].

The electronic polarization amounts to approximately 1 eV and is the largest contribution to the overall polarization. The ion induces a shift of the π-orbitals with respect to the atoms of the molecule. Therefore, the centers of gravity of positive and negative charges do not overlap resulting in a dipole. Due to the redistribution of charge, the ion and the next-neighbor molecules relax in their geometrical structure to minimize the total energy of the system. This effect is called intramolecular polarization and amounts to 150 meV in anthracene. Another important contribution is the lattice polarization, which is a result of a shift of the molecules due to the interaction between the ion and the polarized next-neighbor molecules. The lattice polarization energy is usually smaller than 100 meV and by up to 3 orders of magnitude slower than the electronic and intramolecular polarization. Therefore, it can be neglected on the time scale of lifetimes of excited molecules.

The dependence of the energy levels of an organic solid on the polarization effects is schematically shown in Figure 4.5. The ionization energy (I) and the electron affinity (χ) of a molecule in the gas phase are the minimum energy for the formation of a separated free electron and hole. The removal or addition of an electron from a molecule in an organic solid results in a polarization of the surrounding molecules. Consequently, I and χ are reduced and increased to the solid-state values IP and EA, respectively. The ionization energy of a solid IP is measured by using ultraviolet photoemission spectroscopy (UPS). Here, an electron is removed from a molecule and the resulting hole will induce polarization energy P_+ in the surrounding medium that will shift the HOMO to lower energy.

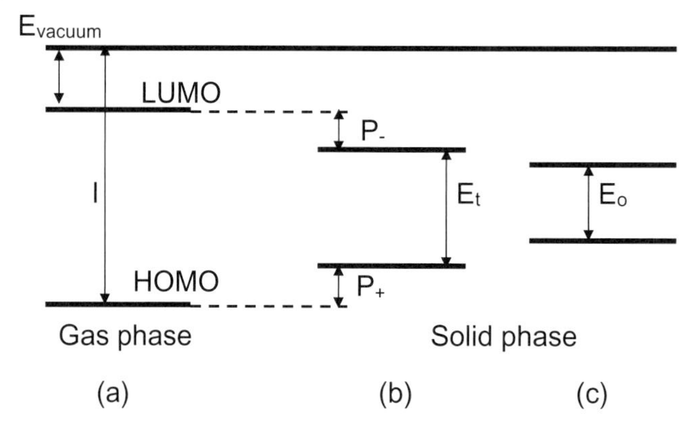

Figure 4.5 Schematic diagram representing the dependence of the polarization effect $P = P_+ + P_-$ on the energy levels in organic solids: (a) molecular ion in gas phase; (b) molecular ion in condensed phase; (c) optical gap of neutral molecule.

The spectrum of the HOMO will include both polarization mechanisms described earlier, that is, the electronic and intramolecular polarization. The situation is vice versa for the LUMO measured by inverse photoemission spectroscopy (IPES). The additional electron added to the LUMO results in the polarization energy P_-. The energy difference between HOMO and LUMO determined by UPS and IPES, respectively, is called the transport gap E_t. Since UPS and IPES are both surface-sensitive techniques, the polarization induced by the molecules on the surface is smaller compared to the polarization in the bulk because of the reduced number of next-neighbor molecules. Therefore, the transport gap determined from the energy separation of the center of mass of the HOMO and LUMO features recorded by UPS and IPES is called "surface transport gap" and is larger than the bulk transport gap. Choosing the peak-to-peak energy difference of the HOMO and the LUMO to determine the transport gap, one has to take into account that compared to the gas phase spectra, vibrational excitations tend to shift the HOMO and the LUMO peak away from the Fermi level. The measured peak-to-peak energy separation increases by the Franck-Condon maximum and can be estimated by a 100-meV shift for each peak. Therefore, 0.2 eV has to be subtracted from the experimentally determined peak-to-peak HOMO–LUMO energy difference.

Hill *et al.* performed combined photoemission and inverse photoemission measurement on PTCDA films and determined the surface transport gap to amount to 4.0 eV. Comparing the spectra of their films with available gas phase UPS data [171], they evaluated polarization energies P_+ of 1.15 ± 0.3 eV. The correction to obtain the bulk from the surface polarization was experimentally estimated by Salaneck to be around 0.3 eV for molecules like anthracene [172]. Assuming that P_- includes in principle the same components as P_+, the surface transport gap has to be decreased by 0.6 eV to take into account the difference in surface and bulk polarization. Subtracting another 0.2 eV for the vibrational excitations, they estimated the bulk transport gap for PTCDA films to be (3.2 ± 0.4) eV [169].

These values are supported by theoretical calculations and experimental investigations for PTCDA from Tsiper *et al.* [173]. They have calculated the electronic polarization energy, that is, $P = P_+ + P_-$, self-consistently in terms of charge redistribution within the molecule. Both experiment and theory show that P is 500 meV larger in a PTCDA monolayer than in 5 nm films. Calculated values for the bulk and the free surface are 1.82 and 1.41 eV, respectively.

The optical gap, on the other hand, is the energy necessary to excite an electron from the HOMO into the LUMO and is defined as the energy corresponding to the maximum of the optical absorption peak. For covalently bonded inorganic semiconductors, energy levels are described in terms of delocalized states. This results in efficient screening and exciton binding energies of only a few milli electron volt. Therefore, E_t values can be approximated by the onset of the optical absorption (E_{opt}). The strong localization of free carriers and the higher polarization energies in organic solids result in the exciton binding energies in the order of ~1 eV. Therefore, the transport gap is found to be considerably larger than the optical one involving exciton formation. Thus, in order to understand electronic and electrical characteristics for organic electronics, it is important to determine the accurate energy levels responsible for carrier transport.

Recently, Hill *et al.* determined the exciton binding energies of organic molecular films, based on results from UPS and IPES measurements [169]. They found that the values strongly depend on the morphology of the films. In the case of crystalline films, the exciton binding energy is reported to be ~0.5 eV, while amorphous films show values larger than 1.0 eV. The accuracy of those values, however, is limited by the poor resolution of IPES of about ~0.4 eV.

4.1 Photoluminescence

In photoluminescence the sample is excited by light with a photon energy larger than the fundamental optical absorption edge. The light emitted by the sample is recorded as a function of photon energy. To measure the lifetime of optically excited states a pulsed laser system and a time resolving photon detector are used.

A typical experimental set-up for photoluminescence spectroscopy is shown in Figure 4.6. For measurements on organic semiconductors light in the visible to UV region has to be used. The pulsed light source shown in Figure 4.6 is a Ar^+ laser combined with a dye laser and a cavity dump. While the gas laser only emits light at certain photon energies, the dye laser offers the possibility to tune the photon energy in a certain energy range. The short light pulses are produced by the cavity dump. A cavity dump uses a variable attenuator, where a high attenuation corresponds to a low quality factor (Q-factor) of the resonator. Switching from low to high Q-factor starts stimulated emission, that is, laser operation. Switching back to low Q-factor dumps the laser light from the cavity all at once. A cavity dump is often accompanied by mode locking.

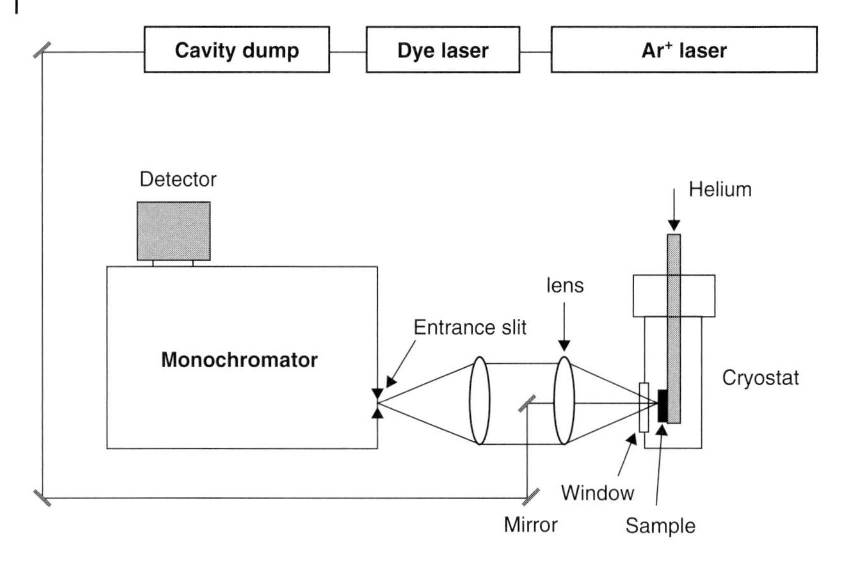

Figure 4.6 Typical photoluminescence setup.

The pulsed laser light is focussed via a lens onto the sample in a spot of lens than 1 mm diameter. To investigate the influence of temperature on the lifetime of optical transitions the sample is mounted on a Kryostat in vacuum. The light emitted from the sample is collect by the same lens and focused onto the entrance slit of a monochromator. Within the monochromator the light is dispersed in energy and project onto the exit slit where a time-resolving photon detector is mounted. Data acquisition of the photon detector is triggered by the pulsed laser, and a delay stage is used to set the time interval between excitation and data acquisition. Plotting the intensity as a function of the time set by the delay stage reveals the lifetime of optically exited states.

4.1.1
PL on PTCDA/Si(111)

The photoluminescence of PTCDA single crystals has been investigated in detail using experimental [174, 175, 176] and theoretical [177, 178, 179], techniques. It has been found that the PL arises from Frenkel excitons (FEs), charge-transfer excitons (CTE), and excimer states. In addition, a further recombination channel called slow (S)-band has been found and whose origin is still not known.

A. Yu. Kobitski and coworkers [180] have investigated the photoluminescence of PTCDA films and compared the results with data taken from PTCDA single crystals. The experimental data and the data evaluation is presented here. Figure 4.7 shows a PL spectrum taken at 35 K for a 40 nm PTCDA film grown on Si(111) by OMBD. Prior to the growth of PTCDA the silicon has been passivated by

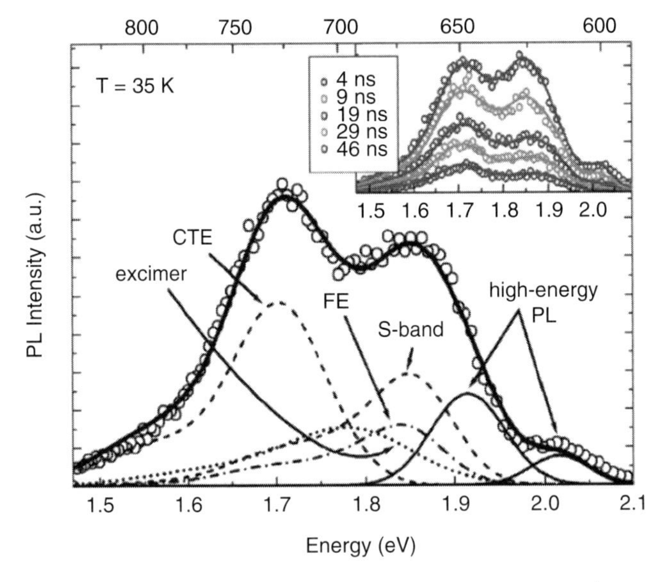

Figure 4.7 The time-integrated PL spectrum of a 40 nm PTCDA film grown on H-passivated Si(111) measured at 35 K. The inset shows the PL spectra for different delay times. Open circles represent experimental data while solid lines shown the results from a detailed data analysis. [from Reference 180]

hydrogen using a wet chemical procedure involving HF. The line shape of the PL spectrum shows at least two distinct maxima indicating that several recombination processes contribute to the photoluminescence. Besides the PL contributions already observed in crystals, that is, charge transfer excitons (CTE), excimer states, Frenkel excitons (FE), and S-band, two high-energy photoluminescence transitions are found. In addition, the inset of Figure 4.7 shows that the overall intensity as well as the intensity distribution within the investigated spectral region changes as a function of the delay time for which the spectrum has been taken. The overall shape of the data shown in Figure 4.7 indicates that a multi-peak fitting procedure has to be employed for a more detailed analysis.

In this fitting procedure each contribution to the photoluminescence spectrum is modelled by a Gaussian function at an energy position ω_j with an energy broadening σ_j and area a_j. The total intensity is the sum of the Gaussian functions multiplied by a factor ω^3 which is related to the density of states of the emitted photons [181]:

$$I_{PL}(\hbar\omega) = \omega^3 \sum \frac{a_j}{\sigma_j \sqrt{2\pi}} \exp\left(-\frac{1}{2}\left(\frac{\omega - \omega_j}{\sigma_j}\right)^2\right) \tag{4.1}$$

As explained in the introduction of Chapter 4 optical transitions couple to the vibronic modes. This results in an additional intensity distribution on the low energy side of each principal photoluminescence transition. In the data analysis presented here this intensity distribution due to vibronic bands is modelled by using additional Gaussian function which accompany the principal transition. Their relative shift in energy, full width at half maximum, and area ratio are ~140 meV, 125–160 meV, and 0.4, respectively.

Curve fittings like the one shown in Figure 4.7 are now performed for various sample temperatures to determine the temperature dependence of the decay time and the PL intensities. Figure 4.8 shows the data for the different optical transitions which contribute to the PL spectrum of the PTCDA film. The solid lines are the result of a simulation of the temperature dependent decay time and PL intensity, where radiative as well as non-radiative processes and activation barriers are taken into account according to: [182]

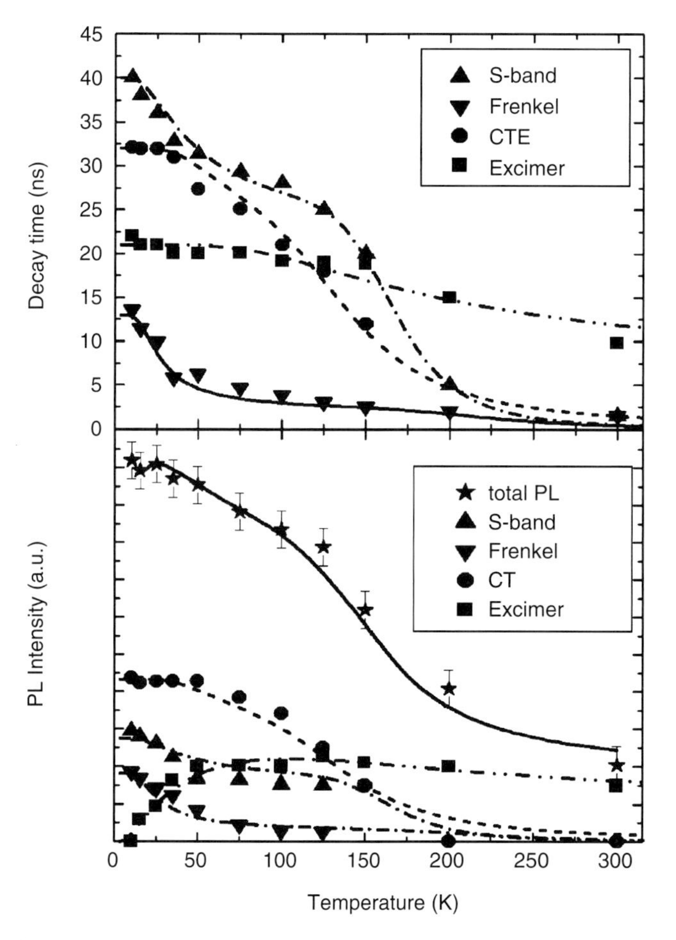

Figure 4.8 Photlumoinescence decay times and intensitiesof a PTCDA film as a function of temperature. Solid lines show the results from the simulations. [from Reference 180]

$$\frac{1}{\tau(T)} = \frac{1}{\tau_{rad}} + \sum \gamma_j^{nrad} \exp\left(-\frac{\Delta_j^{nrad}}{k_B T}\right) \quad (4.2)$$

and

$$I_{PL}(T) = I_0 \exp\left(-\frac{W}{k_B T}\right)\frac{\tau(T)}{\tau_{rad}} \quad (4.3)$$

In (4.2) τ_{rad} is the decay time of the radiative process, while γ_j^{nrad} and Δ_j^{nrad} are the rate and the activation energy of the nonradiative process, respectively. I_0 is proportional to the density of excitons in the initial states after optical excitation, and W the barrier between the initial states after optical excitation and the emissive states.

In all cases except for the excimer PL, the temperature dependence of the PL intensities were just proportional to the decay times, i.e. the barriers W were close to zero. The excimer PL, on the other hand, shows a barrier of about 2.5 meV between the precursor and the excimer state.

The results of the fitting procedure are as follows:

The excimer exciton is found at 1.78 eV with a 2.5 meV barrier between the precursor and the excimer state. It's temperature dependent decay time is described by a radiative recombination with a lifetime of 21 ns and by a thermally activated non-radiative escape over a barrier of 30 meV.

The Frenkel exciton is found at 1.83 eV. The parameters used in the data analysis are similar to the ones used for the single crystal [175] and the broadening of the vibronic bands is based on theoretical and experimental investigations on the vibrational properties of PTCDA according to:

$$\sigma_j^2 = \Delta E^2 = \sigma_{0j}^2 + \alpha_{\text{int}}^2 (\hbar\omega_{\text{int}})^2 \coth\left(\frac{\hbar\omega_{\text{int}}}{2k_B T}\right) + \alpha_{ext}^2 (\hbar\omega_{ext})^2 \coth\left(\frac{\hbar\omega_{ext}}{2k_B T}\right), \quad (4.4)$$

The first temperature dependent term takes into account the contribution of internal (int), that is, molecular vibrations and the values of $\alpha_{\text{int}}^2 = 0.29$ and $\omega_{\text{int}} = 233\,\text{cm}^{-1}$ are for the lowest frequency vibrational mode in a PTCDA molecule. [183] For the external (ext) vibrational modes, or, in other words, phonons, values of $\alpha_{ext}^2 = 7.5$ and $\omega_{\text{ext}} = 50\,\text{cm}^{-1}$ are taken from Raman measurements. [184] The radiative lifetime of the Frenkel exciton amounts to 13 ns with thermally activated non-radiative decay channels.

The S-band emission appears at 1.84 eV and it's radiative lifetime is 40 ns.

Compared to the other photoluminescence contributions the charge transfer exciton at 1.7 eV shows different properties compared to the ones in a PTCDA crystal. A comparison between decay times and PL intensities of the single crystals and films are shown in Figure 4.9. The lifetime and the relative contribution to the total PL intensity are about a factor 2 smaller than for the crystal.

The high-energy PL bands at 1.92 and 2.01 eV are attributed to the recombination from monomer or surface states. Here, they are fitted with a single Gaussian functions with a FWHM of about 100 meV.

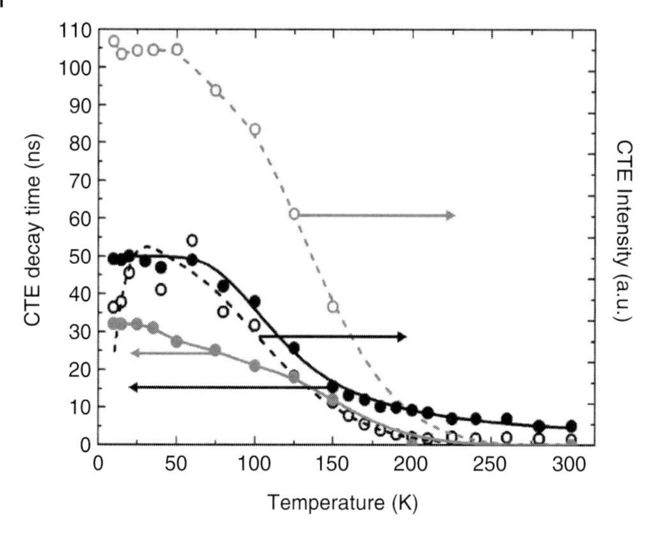

Figure 4.9 Comparison of CTE properties in PTCDA thin films and crystals. Grey and black lines are for thin film and single crystal data, respectively. [from Reference 180]

4.2 Raman Spectroscopy

Raman spectroscopy is based on the inelastic light scattering of light. It measures the energy transferred between an incident photon and the sample. The scattered photon has an energy $\hbar\omega_s$ which is given by

$$\hbar\omega_s = \hbar\omega_i \pm \hbar\Omega \tag{4.5}$$

where $\hbar\omega_i$ is the energy of the incident photon and $\hbar\Omega$ the energy of an elementary excitation in the sample, for example, a phonon, a plasmon, a polariton, a coupled plasmon-phonon, or a single electron or hole excitation. In addition to the energy conservation given by Eq. (4.5). the momentum conservation law gives the relation between the wavevectors $\mathbf{k}_i$ and $\mathbf{k}_s$ of the incident and scattered light, respectively, and the excitation wavevector $\mathbf{q}$:

$$\mathbf{k}_s = \mathbf{k}_i \pm \mathbf{q} \tag{4.6}$$

The "–" sign in Eqs. (4.1) and (4.2) stands for a Stokes process, where an elementary excitation is generated. The "+" sign, on the other hand, stands for an anti-Stokes process where an annihilated excitation is generated. The efficiency of both processes, that is, the ratios of intensities of a certain excitation $\hbar\Omega$ in a Stokes (I_S) and anti-Stokes process (I_{aS}), scales with the temperature of the sample T according to

$$\frac{I_{aS}}{I_S} = \exp\left(-\frac{\hbar\Omega}{k_B T}\right) \tag{4.7}$$

where k_B is the Boltzmann constant.

In linear optics, the interaction between light and matter is described by

$$\mathbf{P} = \varepsilon_0 \hat{\chi}(\omega_i, \omega_s) \mathbf{E}(\omega_i) \tag{4.8}$$

Here, ε_0 is the vacuum dielectric constant and $\mathbf{P}$ is the dipole moment induced by the electric field vector $\mathbf{E}$ of the incident light. The dielectric susceptibility tensor $\hat{\chi}(\omega_i, \omega_s)$ describes the reaction of the solid, that is, how scattered light with frequency ω_s is generated by the incident light with a frequency ω_i. For $\hat{\chi}(\omega_i, \omega_s)$ being constant in time and space, only elastically scattered light exists. In real solids, $\hat{\chi}(\omega_i, \omega_s)$ fluctuates due to elementary excitations like lattice vibrations (phonons). Since the actual relation between the elementary excitations and the susceptibility is not known, the susceptibility is expanded into a Taylor series as a function of the normal coordinate $\mathbf{Q}_j$ of the vibration. $\mathbf{P}$ is then given by

$$\mathbf{P} = \varepsilon_0 \hat{\chi}_0(\omega_i, \omega_s) \mathbf{E}(\omega_i) + \varepsilon_0 \frac{d\hat{\chi}(\omega_i, \omega_s)}{d\mathbf{Q}_j} \mathbf{Q}_j \mathbf{E}(\omega_i) + \varepsilon_0 \frac{d^2\hat{\chi}(\omega_i, \omega_s)}{d\mathbf{Q}_j^2} \mathbf{Q}_j \mathbf{E}(\omega_i) + \cdots \tag{4.9}$$

The first term oscillates with the frequency of the incident light and represents the elastically scattered light. The second term describes first order scattering processes, that is, the excitation of a phonon. The third term describes processes where two elementary excitations are involved. The partial derivatives in (4.9) are termed Raman tensor.

The interaction between the incident photon and the elementary excitations in the sample are mediated by electronic interband transitions. In a simple picture, the incident photon excites an electron–hole pair, where the electron and/or hole interact with the elementary excitations. The recombination of the electron–hole pair results in the emission of light with a shifted frequency. These interband transitions define the dielectric susceptibility in the visible spectral range where Raman experiments are usually performed.

A typical Raman experiment is shown in Figure 4.10. The excitation source is a laser, which gives a well-defined frequency ω_i of the incident light. The laser light is guided by a set of mirrors on to the sample. In this particular setup, the sample is an UHV chamber and the laser light is focused with a lens through an UHV window on the sample. The scattered light is collected through a second window using an objective lens. While the scattered light is detected in a direction perpendicular to the sample surface, the incident light is inclined by a certain angle. Since Eq. (4.6) describes the relation between wavevectors inside the sample and the refractive index of most samples is reasonably high, k_i is antiparallel to k_s and this geometry is called back-scattering geometry. The objective lens focuses the scattered light onto the entrance slit of a monochromator. Since Raman signals are usually about six orders of magnitude smaller than the elastically scattered light, these monochromators include a band-pass to decrease the background intensity. After passing the band pass, the selected region of the spectrum is dispersed by a monochromator. A multichannel detector located at the exit slit position of the monochromator records the spectrum.

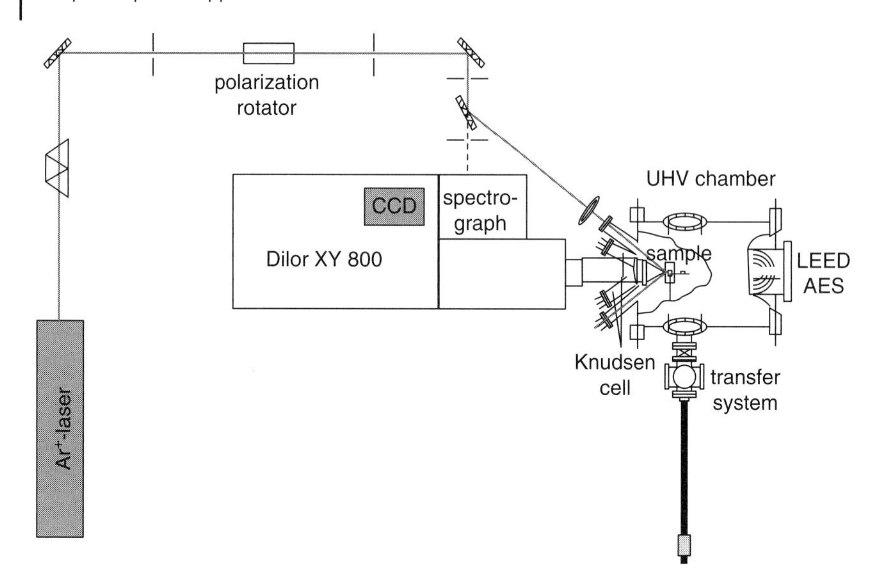

Figure 4.10 Raman setup with Ar⁺-laser, UHV chamber for in situ studies of organic thin film growth, monochromator with multichannel CCD detector.

In the following, Raman investigations on PTCDA will be discussed. The Raman measurements were performed using the 2.54 eV emission line of an Ar^+ laser for excitation. This laser line was chosen because it is resonant with the first maximum of the absorption spectra of both molecules shown in Figure 4.4. The films were measured in a backscattering geometry with a spectral resolution of 2.5 cm^{-1}. The incident beam, having a power of 30 mW, was focused onto a spot of ~300 μm in diameter. For comparison, Raman measurements were taken from a crystal obtained by sublimation using an Olympus microscope with 100× magnification objective in backscattering geometry with a spectral resolution of 1.2 cm^{-1}. The spot size in this case was ~1 μm in diameter and the power measured in the focus was 0.08 mW.

4.2.1
Raman Spectroscopy on PTCDA Crystals and Thin Films

Two polarization configurations were used for each sample position. In the Porto notation, $z(xx)\underline{z}$ and $z(xy)\underline{z}$ denote the cases where the electric field vector of scattered light is parallel/perpendicular to that of the incident light (parallel/crossed polarization configuration). The position of the polarization analyzer was maintained fixed in all experiments, in order to eliminate the systematic error, which might be induced by the different response of the spectrometer to differently polarized light. For the measurements performed on the molecular films, the laboratory axes coincide with the substrate axis ($x = x_s = [110]$, $y = y_s = [110]$, $z = z_s = [100]$) when the angle of rotation around the surface normal is $\gamma = 0°$. The orientation of molecules is determined from the depolarization ratio of the breathing mode at ~220 cm^{-1} obtained upon rotation of the sample around the surface normal.

For a simulation of the depolarization ratio, the Raman tensor specific for the A_g modes is determined from density functional calculations using Gaussian'98 package at the B3LYP level of theory with a standard 6-31G(d) basis set [166]. The symmetry of the HOMO of DiMe-PTCDI is A_u, and that of the LUMO is B_g. Therefore, a resonant Raman effect that involves a HOMO–LUMO transition couples vibrational transitions with an electronic transition dipole of which is oriented along the x_m axis (where x_m denotes the long axis within the molecular plane, while y_m and z_m denote short axis and the direction perpendicular to the molecular plane, respectively). Under resonant excitation, the off-diagonal components disappear and the Raman tensor reads

$$A_g^m = \begin{pmatrix} 1 & 0 & 0 \\ 0 & 0.04 & 0 \\ 0 & 0 & 0 \end{pmatrix} \tag{4.10}$$

for the lowest frequency internal A_g mode at 221 cm^{-1}.

In Figure 4.11, Raman spectra of a DiMe-PTCDI single crystal and a DiMe-PTCDI thin film are shown for the two different polarization geometries. For molecules with a center of inversion vibrations that are symmetric with respect to

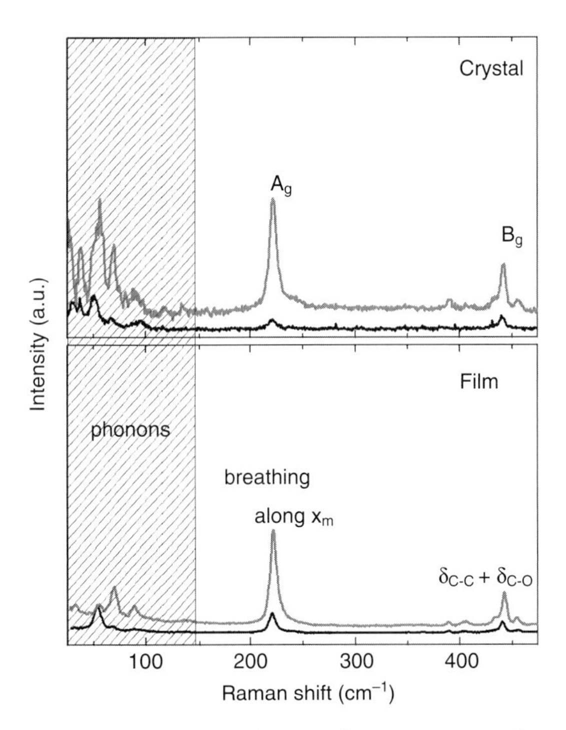

Figure 4.11 Comparison between the Raman spectra of DiMe-PTCDI crystal and film obtained in $z(xx)\underline{z}$ (upper red curves) and $z(xy)\underline{z}$ (lower black curves) polarization configurations.

it are labeled with a g and are usually Raman-active. The antisymmetric vibrations are labeled with u and are infrared-active. Taking spectra under resonance conditions enhances predominantly the totally symmetric A_g modes. The A_g mode at $221\,cm^{-1}$ is a breathing mode, while the one at $441\,cm^{-1}$ has out-of-plane character involving deformation of C–C and C=O bonds.

As a consequence of crystal formation with two molecules per unit cell, external vibrational modes can be observed in the Raman spectra. The vibrational representation for the external modes is $\Gamma = 3A_g + 3B_g + 2A_u + B_u$. The six modes having even symmetry will show Raman activity at frequencies below $125\,cm^{-1}$. As seen in Eq. (4.10), phonons appear in the Raman spectra of single-phase crystals as well as in those of films exhibiting similar polarization response. The phonon intensity, relative to that of the internal modes, is larger in the crystal compared to film spectra. The difference can be explained by the larger size of the crystals compared to that of the grains in the films. Moreover, the spot size in the case of films allows probing of more than 10^5 grains with various sizes. The strong polarization response for both external and internal modes is remarkable and indicates a preferred orientation of the grains with respect to the substrate axes.

The packing of the molecules into a crystalline environment is expected to affect the internal modes due to in-phase and out-of-phase coupling. This effect called *Davydov splitting* depends on the number of molecules in the unit cell and their dipole and quadrupole interaction. The former determines the multiplicity and the latter the amount of splitting. For DiMe-PTCDI with two molecules in the unit cell, there should be a twofold splitting. Due to the fact that the angle between the molecules in the unit cell is ~36°, the dipole interaction which, is proportional to the cosine of that angle, is so small that the amount of splitting lies at the resolution limit of the Raman experiment. Therefore, the use of a single-molecule Raman tensor is likely to be justified in this particular case.

The spectra obtained upon sample rotation are shown in Figure 4.12. For a quantitative analysis of the polarization response one defines the depolarization ratio as the intensity ratio between the Raman signal obtained in crossed polarization configuration to that obtained in parallel configuration. The experimental depolarization ratio of the $221\,cm^{-1}$ mode for a film is depicted as a function of rotation angle γ. In order to extract the geometrical arrangement of the molecules from the depolarization ratios there are three coordinate systems to take into account: molecular (x_m, y_m, z_m), substrate (x_s, y_s, z_s), and laboratory (x, y, z).

The laboratory coordinate system (x, y, z) was defined such that the z-axis is parallel with the direction of the incident beam and perpendicular to the sample surface. The x-axis was defined to be parallel to the direction of the polarization analyzer. The electric-field vector of the incident radiation was either parallel or perpendicular to the x-axis (parallel to the y-axis). The laboratory axes coincide with the substrate axes ($x = x_s = [011]$, $y = y_s = [0\text{–}11]$, $z = z_s = [100]$) when $\gamma = 0°$. Two polarization configurations were used for each sample position. In the Porto notation, $z(xx)\underline{z}$ and $z(yx)\underline{z}$ denote the cases when the incident electric field vector of scattered light is parallel/perpendicular to that of the analyzed light (parallel/crossed polarization configuration).

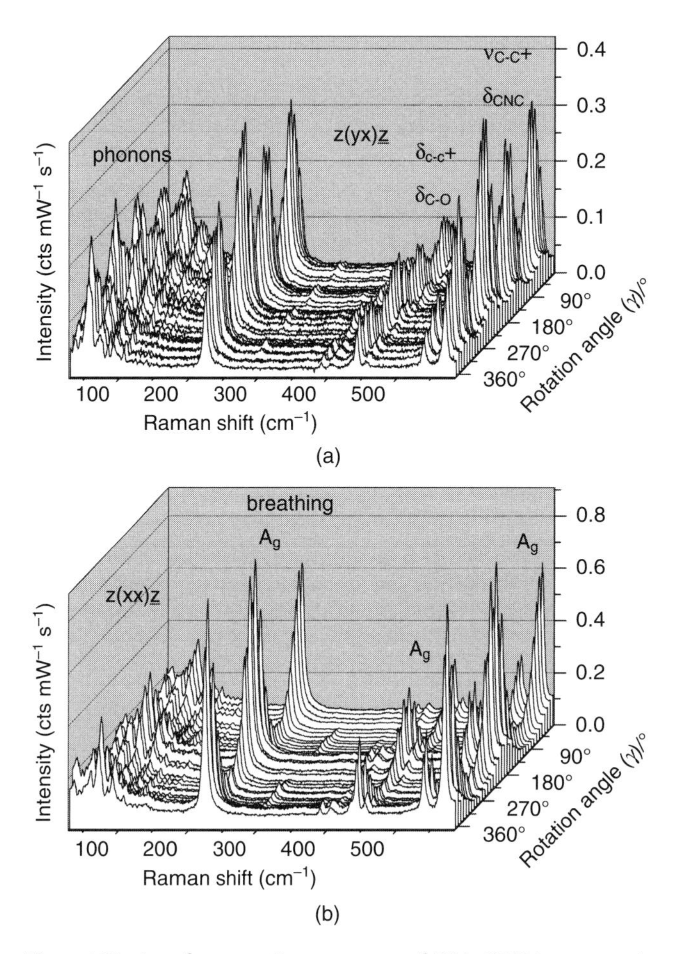

Figure 4.12 Low-frequency Raman spectra of DiMe-PTCDI upon rotation of the sample around its normal (with angle γ). Upper (a) and lower spectra (b) are recorded in crossed and parallel polarization, respectively.

Raman intensities of A_g modes vary with a period of 180° and 90° in parallel and in crossed polarization, respectively. The maximum response in parallel polarization for the A_g modes takes place when the electric field vectors are parallel to the [011] direction of the substrate ($\gamma = 0°$, 180°). This indicates a good ordering of the molecules with a preferred orientation of their x_m-axis close to the [011] substrate axis, in agreement with the NEXAFS results.

The experimental depolarization ratio of the 221 cm^{-1} mode is shown as a function of rotation angle γ by symbols in Figure 4.13. Two consecutive transformations are required to transform the molecular Raman tensor to the Raman tensor $A_{g,l}$ in the laboratory reference frame. The first orthogonal transformation can be applied using the Eulerian angles (φ, θ, ψ), which were previously applied by Aroca

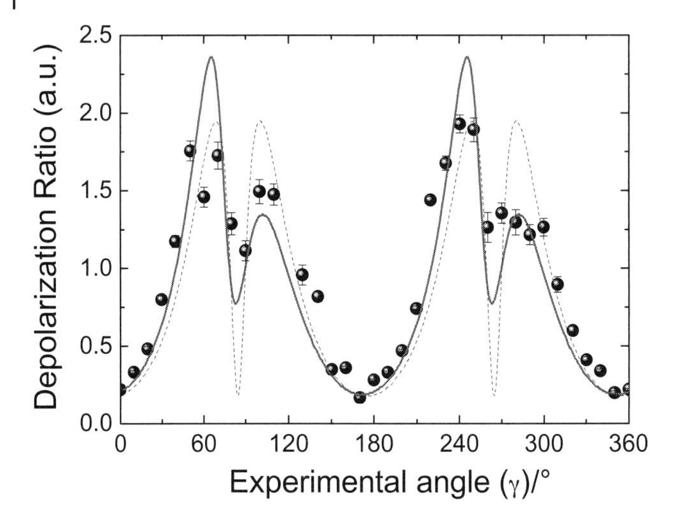

Figure 4.13 Experimental (symbols) and simulated (lines) depolarization ratios of the breathing mode at ~221 cm^{-1} obtained upon rotation around sample normal with the angle γ for a DiMe-PTCDI film.

et al. to Raman study of orientations in a highly symmetric molecular system [184]. The second transformation is from the substrate to the laboratory coordinate system and implies a clockwise rotation around the substrate normal (z_s) with the angle γ. The Raman intensity is then calculated as $I = (\mathbf{e}_s \cdot A_{g,l} \cdot \mathbf{e}_i)^2$, where $\mathbf{e}_i$ and $\mathbf{e}_s$ are the electric field vectors for the incident and the scattered light, respectively.

A least-square fit of the experimental depolarization ratio was performed using a Levenberg-Marquard algorithm. The closest match between the calculated depolarization ratios using the one-molecule approximation is presented in Figure 4.13 by the dashed line. The corresponding set of Euler angles is ($\varphi = -70°$, $\theta = 53°$, $\psi = 51°$), which means that the molecular plane is tilted with respect to that of the substrate by $53° \pm 5°$, and the angle between the projection of x_m onto the substrate plane and [011] substrate axis is $-7° \pm 5°$. This set of angles is very close to those determined from NEXAFS. Even though the main maxima of the experimental data are reproduced well, the steep minima and the same height of all maxima indicate that a more complex model is required.

Considering the crystalline nature of the film reflected by the presence of phonons, a natural approach would be to consider two noninteracting molecules, the Raman signals of which add up. The model was constructed such that the angles between the two molecules in the unit cell are maintained close to the angles reported for the single crystal and only the unit cell is rotated with respect to the substrate by changing the Euler angles. The coordinate system of the unit cell (x_u, y_u, z_u) was chosen so that the x_u and y_u axes are contained in the (102) crystallographic plane. The molecular planes of the two molecules are tilted with

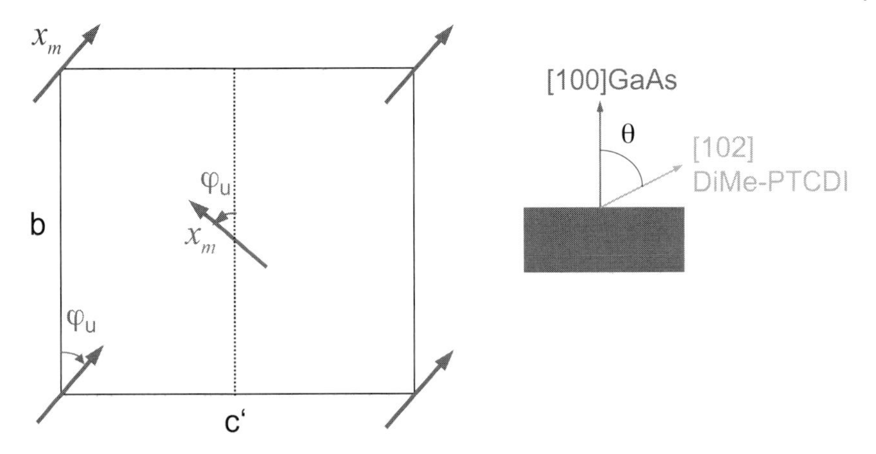

Figure 4.14 Schematic of the DiMe-PTCDI unit cell in the (102) crystalline plane (left) and of the relative orientation of the (102) planes with respect to the S-passivated GaAs(100) substrate (right, both defined by the normal direction).

+8° and −8° with respect to the (x_u, y_u) plane, and the long axis of the two molecules are rotated by +18° and −18° with respect to the x_u-axis.

The lowest deviation between the calculated and the experimental data is provided by the following set of Euler angles: $\varphi = 114° \pm 5°$; $\theta = 56° \pm 4°$, $\psi = 28° \pm 5°$. This means that the crystal (102) plane forms an angle of 56° with the substrate plane, and the projections of the long molecular axis deviate from the [011] direction of the substrate by $\chi_1 = -7° \pm 5°$ and $\chi_2 = -48° \pm 11°$, respectively. Figure 4.14 shows a sketch of the unit cell in the (102) DiMe-PTCDI crystal plane and their relative orientation with respect to the GaAs substrate.

It must be noted that in each of the two models, the minima of the depolarization ratio should be very close to zero assuming that the scattered light is totally polarized. In order to reproduce the experimental values in the minima, the scattered intensities in parallel and perpendicular configurations had to be mixed by a constant factor D:

$$\mathrm{Dep} = \frac{(1-D)\times I_{yx} + D\times I_{xx}}{D\times I_{yx} + (1-D)\times I_{xx}} \tag{4.11}$$

The parameter D was also optimized during the fitting procedure and for both models, the resulting value was $D \sim 0.15$. This parameter describes a depolarization in the light scattered by the sample due to the surface roughness [185, 186], thickness nonuniformity [185], angular spread of the collected beam [185, 186], and eventually due to a spread in the orientation of some molecules from the preferential arrangement.

Raman spectroscopy may also be used to investigate the interaction between metals and perylene derivatives. Figure 4.15 shows the Raman spectra of bare PTCDA and DiMe-PTCDI films and after deposition of 43 nm of indium or silver.

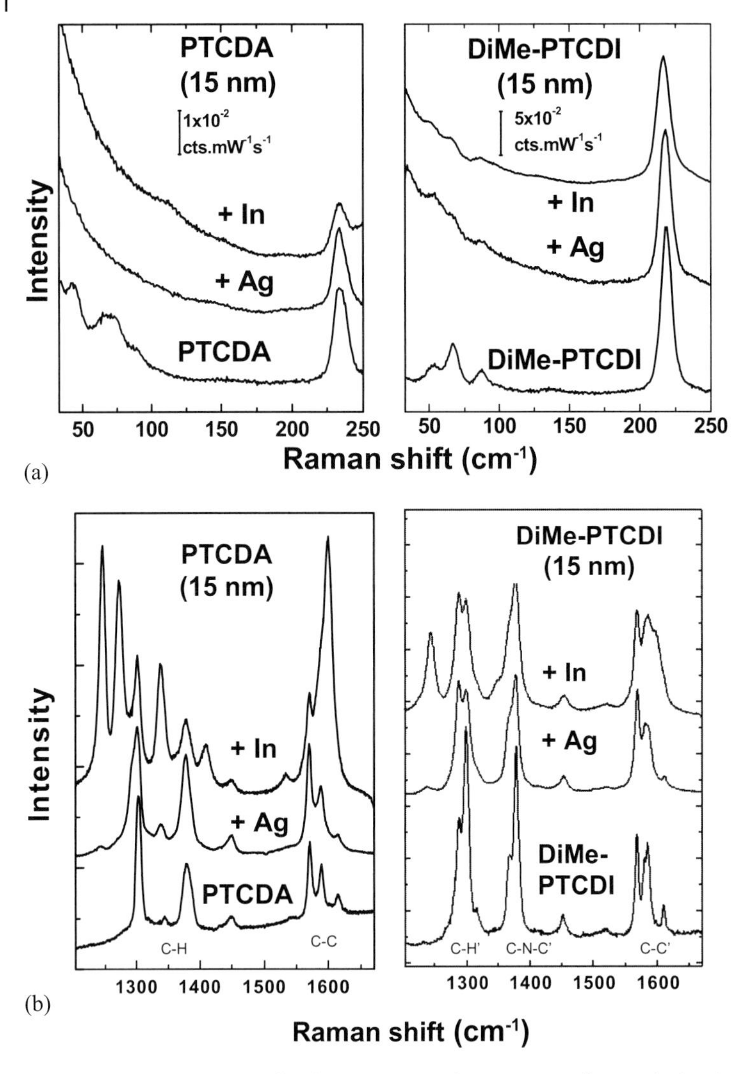

Figure 4.15 Raman spectra for clean PTCDA and DiMe-PTCDI films and after deposition of In or Ag.

It should be mentioned that the deposition of metals results in an enhancement of the Raman features, an effect also known as surface-enhanced Raman scattering (SERS). Two mechanisms are responsible for the intensity enhancement. One mechanism is a chemical enhancement via the increase in polarizability of the Raman scatterer due to electronic interaction between the adsorbate and the metal. This effect is confined to the first layer. The other mechanism is an enhancement of the electric field in the vicinity of the metal. A detailed presentation of using SERS to investigate the growth of Ag on PTCDA and DiMe-PTCDI can be found in Reference [187]. Figure 4.15a shows the frequency region between 25 and 250 cm^{-1}. The most dominant feature in this spectral region is the breathing mode at 233 and 221 cm^{-1} for PTCDA and DiMe-PTCDI, respectively. The lower frequency in the case of DiMe-PTCDI can be attributed to the larger mass of this molecule resulting from the methyl side groups. Assuming that both molecules can be described in a good approximation as linear oscillators, the ratio of the frequencies of the breathing modes should then depend on the ratio of the molecular masses as

$$\frac{\omega_{\text{DiMe-PTCDI}}}{\omega_{\text{PTCDA}}} = \sqrt{\frac{m_{\text{PTCDA}}}{m_{\text{DiMe-PTCDI}}}} = \sqrt{\frac{392\ \text{amu}}{418\ \text{amu}}} = 0.97 \tag{4.12}$$

The value obtained by using the molecular masses of the molecules is close to the experimental frequency ratio of 0.95. Any change in the chemical structure of the molecules due to bonding of additional atoms will result in a change of frequency of this vibration. The spectra in Figure 4.15a show no change in frequency of this mode due to the deposition of Ag and In. Therefore, any formation of covalent bonds between the metals and the molecules can be excluded. Especially the formation of In_4-PTCDA proposed by Kera *et al.* can be excluded.

At low frequencies, the background is increasing after metal deposition. Therefore, features below 120 cm^{-1}, which are due to the excitation of phonons in the organic films, are difficult to detect. The increase in the background can be attributed to an increase in the surface roughness due to the deposition of the metals. This indicates the formation of metal clusters on top of the organic crystallites and is supported by the XRD results. Due to the formation of metal clusters, Raman spectra can be observed even after the deposition of a metal layer with a thickness of 43 nm.

The high-frequency region presented in Figure 4.15b presents the most dominant internal modes with C–H and C–C character. Deposition of Ag on PTCDA leads to the appearance of a B_{1u} band at 1243 cm^{-1} and an increase in the relative intensity of the B_{3g} mode at 1338 cm^{-1}. These modes are intrinsic modes from molecules having a direct contact with Ag atoms and appear due to the breakdown of selection rules. This breakdown of selection rules is a result of a dynamical charge transfer modulated by the molecular vibrations. Since the total symmetric A_g modes still dominate the spectra, it can be concluded that only a few PTCDA molecules are in contact with the Ag clusters. After the deposition of In, the breakdown of selection rules is even more obvious since B_{3g}, B_{1u}, and B_{3u} modes appear with intensities comparable to the A_g modes. Therefore, more molecules are now in contact with In, which can only be explained by a diffusion of In into the PTCDA crystallites.

As for the deposition of Ag on PTCDA, the deposition of Ag on DiMe-PTCDI induces only small changes in the spectra. The intensity ratio of the C–H modes at 1290 and 1300 cm^{-1} is changed and the two modes at 1369 and at 1380 cm^{-1} merge into one band at 1375 cm^{-1}. The intensity of the B_u mode with C–H character at 1248 cm^{-1} is slightly enhanced, which is again attributed to a breakdown of selection rules as a result of a dynamical charge transfer modulated by the molecular vibrations. This mode is further enhanced in intensity when In is deposited. Again, In seems to diffuse into the organic film, but to a much lesser degree than in PTCDA. The strongest interaction is found for the In/PTCDA interface, which is in agreement with the results from photoemission spectroscopy investigations.

4.3
Infrared Spectroscopy

In infrared spectroscopy the light interacting with matter is of the same frequency as the elementary excitations. Here, a continuous light source is used and the change in intensity as a function of wavelength is measured after transmission through or reflection from a sample. The experimental approach to measure light intensities in the infrared region is different from techniques used in the visible and UV spectral range, that is, the use of grating monochromator. In infrared spectroscopy a technique called Fourier transformed infrared spectroscopy (FTIR) is used nowadays almost exclusively. The layout of such a spectrometer is shown in Figure 4.16.

The light of a broadband source is divided by a beam splitter into two beams. In Figure 4.16, a fixed mirror reflects the reflected beam, while a mirror moving periodically back and forth reflects the transmitted beam. After reflection at the respective mirrors, the two beams are recombined by the beam splitter and focused onto the sample. In the sample stage, the light can be transmitted through a sample or reflected from a sample surface. The transmitted/reflected light is focused on a detector. Behind the beam splitter, the two partial beams

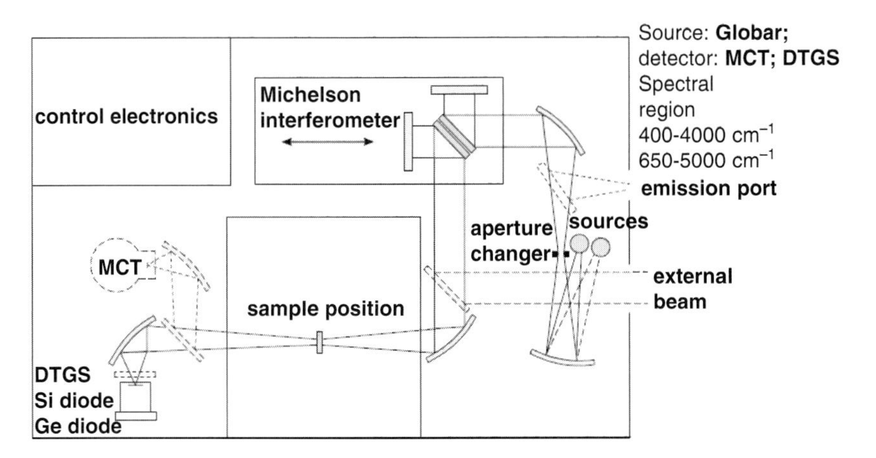

Figure 4.16 Layout of an FTIR spectrometer.

interfere, and the total intensity depends on the phase shift between the two waves for different position of the movable mirror. On the detector, the radiation field is superimposed with a time-delayed copy of itself. Therefore, the detector measures basically the autocorrelation function of the radiation field, which is called the interferogram in FTIR spectroscopy. The Fourier transformation of this autocorrelation function is the desired power spectrum in the frequency domain.

Instead of using the internal sample stage, the light can also be coupled by mirrors and IR transparent windows onto a sample in UHV chamber. The reflected light is then detected via a second IR transparent window and a detector. In addition, polarizers are used for polarization-dependent measurements. The typical standard frequency range is 10–5000 cm^{-1}, but using special detectors measurements can be performed up to 40 000 cm^{-1}.

Samples with a metallic substrate are most suitable for investigations with FTIR, since metals have a high reflectivity in the infrared frequency range. Here, organic thin films with coverages below one monolayer can be investigated. Semiconductors, on the other hand, are transparent in the infrared due to their band gap. The intensity of light reflected back on the detector is, therefore, much lower.

4.3.1 Assignment of Vibrational Modes

For quantitative analysis of the polarized response recorded in IR and Raman spectroscopies, the knowledge of the vibrational mode character and their symmetry is essential. In order to make an assignment of Raman and IR features to vibrational modes, density functional calculations have been carried out using the Gaussian'98 package [30] at the B3LYP level of theory using the standard 3-21G basis set. The point symmetry group of the PTCDA molecule is D_{2h}. PTCDA has 38 atoms and hence exhibits 108 internal modes. Those that are symmetric with respect to the center of inversion (labeled with g) are usually Raman active and those that are antisymmetric (labeled with u) show IR activity. The representation of the internal modes is

$$\Gamma_{\text{PTCDA}} = 19A_g + 18B_{1g} + 10B_{2g} + 7B_{3g} + 10B_{1u} + 18B_{2u} + 18B_{3u} + 8A_u$$

DiM-PTCDI consists of 46 atoms, the point symmetry group of the molecule being C_{2h} or C_{2v}, for the geometry with an inversion center and that without, respectively. There are 132 internal molecular vibrational modes. In the case of C_{2h} symmetry, 66 of them are Raman active ($44A_g + 22B_g$) and 66 are IR active ($23A_u + 43B_u$). For the C_{2h} symmetry, there are $44A_g + 22B_g + 23A_u + 43B_u$ irreducible representations, with the corresponding vibrational modes being either Raman or IR active. Due to the similarity in the calculated frequencies of modes for the two point groups, the geometry with C_{2h} symmetry will be considered for the ease of comparison with PTCDA. Table 4.1. shows experimental and theoretical vibration frequencies for PTCDA and Dime-PTCDI.

Table 4.1 Comparison between the experimental frequencies of the IR modes for PTCDA and DiMePTCDI films on S–GaAs(100) and calculated properties for the corresponding isolated molecules.

PTCDA	DiMePTCDI	PTCDA	DiMePTCDI	PTCDA	DiMePTCDI	PTCDA	DiMePTCDI
733	743	764	786	$B_{3u}(z)$	$A_u(z)$	$^{\delta}$O=C–C + $^{\delta}$C–O–C, oop	$^{\delta}$C–H + $^{\delta}$C–C–C, opp
809	809	853	859	$B_{3u}(z)$	$A_u(z)$	$^{\delta}$C–H + $^{\delta}$C–C–C, opp	$^{\delta}CH_3$ + $^{\delta}$C–C–C, opp
939		947		$B_{1u}(y)$		$^{\upsilon}$C–C + $^{\upsilon}$C–O	
1017, 1024	1022, 1053	1040	1032	$B_{1u}(y)$	$B_u(x)$	$^{\upsilon}$C–O + $^{\upsilon}$CC	$^{\delta}$C–H + $^{\upsilon}$ring + $^{\delta}CH_3$
1236	1237	1256	1260, 1265	$B_{1u}(y)$	$B_u(y)$; $B_u(x)$	$^{\upsilon}$C–O + $^{\upsilon}$C–C	$^{\delta}$C–H + $^{\delta}$ring + $^{\delta}CH_3$;
							$^{\delta}$C–H + $^{\delta}$C–N–C
1300	1285	1309	1317	$B_{2u}(x)$	$B_u(y)$	$^{\upsilon}$C–O + $^{\upsilon}$C–C	$^{\upsilon}$C–O + $^{\upsilon}$C–N–C + $^{\delta}CH^3$
	1350, 1358		1346		$B_u(x)$		$^{\upsilon}$C–O + $^{\upsilon}$C–N–C + $^{\delta}$C–H + $^{\delta}CH_3$
1407	1400	1439	1372	$B_{2u}(x)$	$B_u(x)$	$^{\upsilon}$C–O + $^{\upsilon}$C–C	$^{\delta}$C–H + $^{\delta}CH_3$ + $^{\upsilon}$C–N
					$B_u(x)$;		$^{\delta}$C–H + $^{\upsilon}$C–C;
	1436, 1449		1438, 1456;		$B_u(x)$,		$^{\delta}$C–H + $^{\delta}$ring + $^{\delta}CH3_2$
			1483		$B_u(x)$		CH_3 umbrella def.
1594	1577, 1593	1618	1617	$B_{2u}(x)$	$B_u(x)$	$^{\upsilon}$C–C + $^{\delta}$C–H	$^{\upsilon}$C–C + $^{\delta}$C–H
1731, 1743	1658, 1665	1756	1678	$B_{1u}(y)$	$B_u(y)$	$^{\upsilon}$C=O	$^{\upsilon}$C=O
1771, 1778	1692, 1696	1796	1715	$B_{2u}(x)$	$B_u(x)$	$^{\upsilon}$C=O	$^{\upsilon}$C=O + $^{\delta}CH_3$

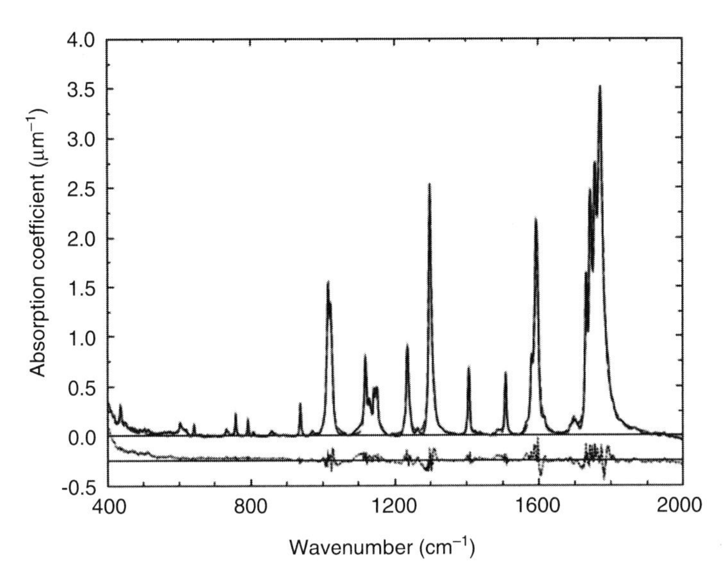

Figure 4.17 Absorption coefficient of a 220 ± 10 nm PTCDA film grown by OMBD on a H-passivated Si(111) substrate. Thick solid curve: experimental spectra; dashed curve: superimposed fit, compare Table 4.1. Dotted line: fit residuum, shifted for clarity to a reference of −0.25 μm^{-1}. At low wavenumbers, the experimental data are influenced by the tail of a broad structure below 400 cm^{-1}. In order to make this contribution more evident, it is not removed from the fit residuum.

4.3.2
IR on PTCDA/Si

The IR transmission of PTCDA films grown on Si by OMBD was measured at normal incidence and the absorption coefficient α reported in Figure 4.17 was calculated from the transmission ratio of the PTCDA-covered Si substrate T_{PTCDA} and of an uncovered reference sample T_{Si}. For the determination of the PTCDA film thickness, we used ellipsometry in the visible [162], giving a result of 220 ± 10 nm. The calculated curve superimposed on the measured spectra is based on the dielectric function discussed in Reference [188].

Compared to powder spectra, the absorption coefficient of the film deposited with OMBD in Figure 4.17 shows all in-plane features, while the relative size of the out-of-plane features is drastically decreased. The low absorbance of the out-of-plane features supports an arrangement of the molecular planes nearly parallel to the substrate surface in OMBD films. In order to further confirm this assumption, additional reflectivity measurements at various angles of incidence were performed, and the results for p-polarized light are displayed in Figure 4.18. While spectra recorded using s-polarized light (not shown) exhibit only strong in-plane features, the spectra measured in p-polarized light reveal the out-of-plane features since, for increasing angle of incidence, the electric field acquires a component

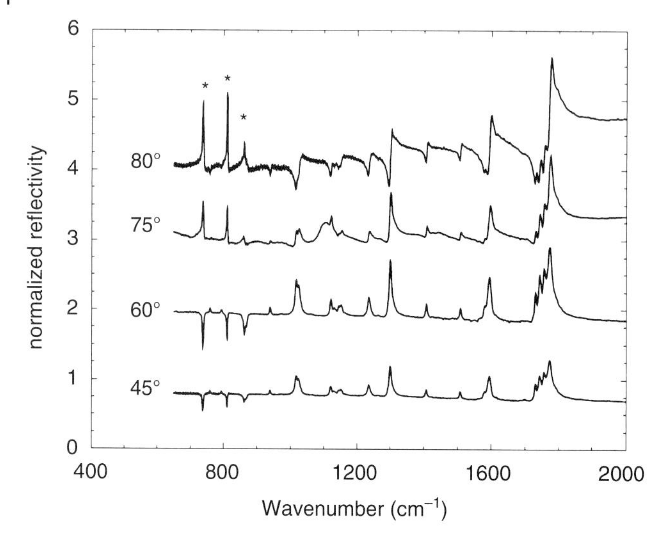

Figure 4.18 Reflectivity of a 220 ± 10 nm thick PTCDA film grown by OMBD on an Hpassivated Si(111) substrate, normalized by the substrate reflectivity. The spectra are measured in p-polarized light for different angles of incidence, and the curves for 60°, 75°, and 80° are shifted for clarity. The film is the same as the one used in Figure 4.17. Out-of-plane modes are marked with *.

along the surface normal. The out-of-plane modes (marked by *) become more prominent for increasing angle of incidence before changing sign at the magic angle. The broad structure around 1100 cm^{-1} in the spectra at 75° is related to oxygen impurities in the substrate.

4.3.3
IR on DiMe-PTCDI

The results of IR measurements for 120 nm DiMe-PTCDI on S–GaAs(100) are given in Figures 4.19 and 4.20. Figure 4.19 shows the IR reflectance spectra measured in s-polarization at 20° angle of incidence with the [011] (thick curves) and [0$\bar{1}$1] (thin curves) directions of the substrate parallel to the plane of light incidence; that is, the electric field is parallel to the [0$\bar{1}$1] and [011] directions. For this near-normal incidence, it is obvious that the s-polarized spectra in [011] and [0$\bar{1}$1] directions show totally different absorption features. From these spectra, we can again derive without any calculation that the film has a strong in-plane anisotropy.

By comparing the experimental frequency and intensity with calculated ones (see Table 4.2), the strongest features measured in the [0$\bar{1}$1] direction can be assigned to vibrations along the x-direction of the molecule. The contribution of features related to other molecular directions is small. Dominance of features related to vibrations in y- and z-directions of the molecule, on the other hand, is observed in the [011] direction spectra. The characteristic features used for further evaluation are marked with x, y, or z.

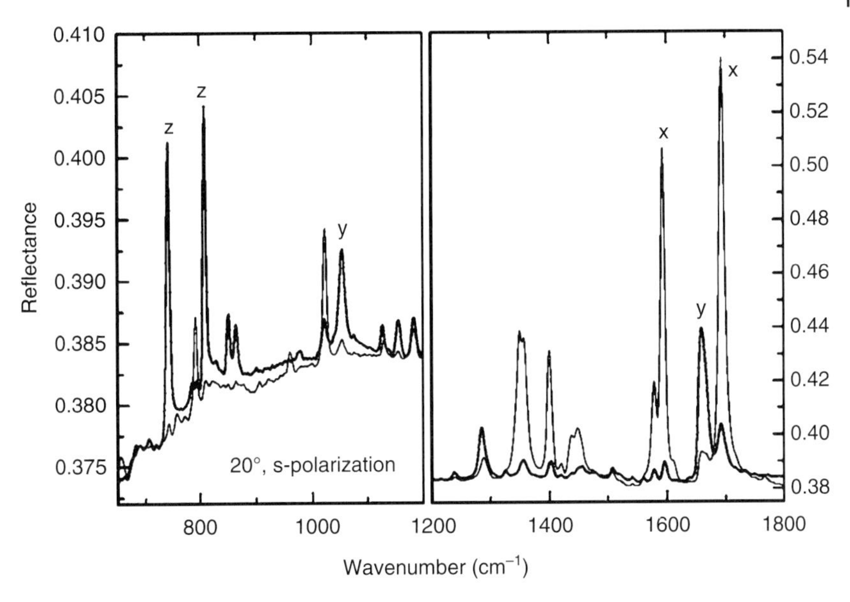

Figure 4.19 Infrared reflection spectra of DiMePTCDI with the [011] direction (thick curve) and [0¯11] direction (thin curve) parallel to the plane of incidence.

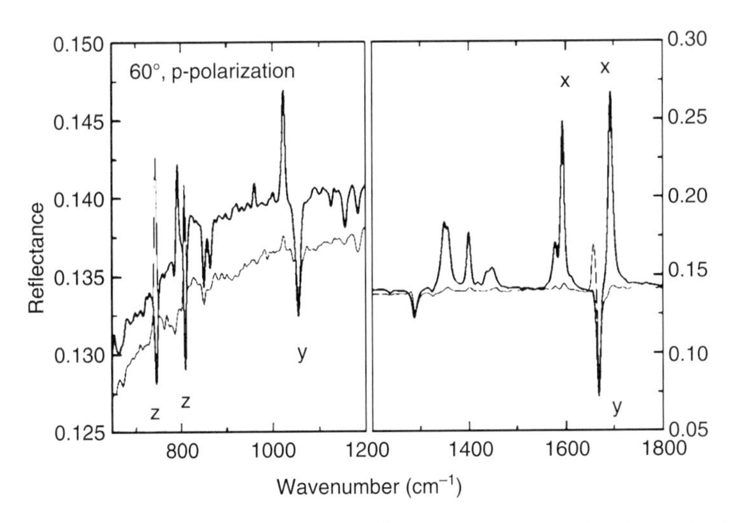

Figure 4.20 Infrared reflection spectra of DiMePTCDI measured at 60° angle of incidence with p-polarized light.

From the spectra in Figure 4.19, we can derive that the DiMe-PTCDI molecules are preferentially arranged with their x-axis in the [011] direction, the direction of the electric field vector, while measuring in s-polarization when the $[0\bar{1}1]$ direction lies in the plane of incidence. The measurements at 60° angle of incidence for s-polarized light (not shown) are similar to those in Figure 4.19 and confirm the in-plane anisotropy.

Table 4.2 Dichroic ratios observed for selected IR modes of DiMePTCDI.

Wavenumber (cm^{-1})	Character	Dichroic ratio	
		I_a/I_b	I_b/I_a
743	z	15.8	
809	z	16.5	
1054	y	4.5	
1596	x		18.1
1658	y	4.7	
1692	x		10

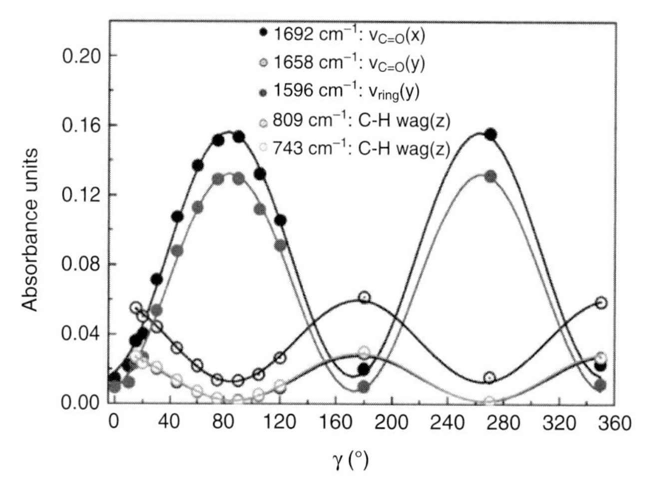

Figure 4.21 Intensities of x, y, and z features of DiMePTCDI as a function of azimuth angle for near normal incidence. Measurements are done with s polarized light.

Additional information about the vibrational modes in the [100] direction can be extracted from the 60° measurements using p-polarized light. The spectra are presented in Figure 4.20. The electric field vector lies in the incidence plane. For the measurement in the [011] direction (thick curve), we see as expected the x features pointing up and additional y and z features pointing down, including information about the [100] direction with respect to the substrate.

In the $[0\bar{1}1]$ direction (thin curve), the contribution of x features is negligibly small. Only z features pointing up and y features pointing down are perceptible. The p-polarized spectra in Figure 4.20 clearly indicate that the molecular planes are tilted with respect to the substrate surface.

Figure 4.21 shows the results of measurements performed at near normal incidence turning the sample around its surface normal by an angle γ starting with

Table 4.3 Spectral position of PTCDA features in the case of both RAS and SE.

Energy (eV)	
RAS	SE
2.15	2.21
2.29	
2.37	–
2.42	2.47
2.54	2.57
2.63	–
–	2.38
–	3.52

the plane of incidence containing the [011] direction of the substrate. The behavior of the DiMe-PTCDI bands can be well described by a $\cos^2$ function as shown in Figure 4.21. The intensity maxima of y and z features are shifted by an angle of about 90° with respect to those of the x features. Maxima are observed at angles $\gamma_x = 7°$ and $\gamma_y = \gamma_z = 94°$. The ratios of the maximum and minimum in-plane intensity values are presented in Table 4.3. With the minimum I_{ax} and maximum values I_{bx} of x features, a coarse evaluation can be performed using the formula given in Reference [189] for the determination of the dichroic ratio D_x. The formula derived for the transmission spectra is applied here to the reflection spectra of a 120-nm-thick film showing absorption features similar to those, which are characteristic for transmission spectra. We obtain

$$D_x = \frac{I_{bx}}{I_{ax}} = \frac{n_a}{n_b} \cot^2 \chi \tag{4.13}$$

Here the refractive indices $n_b = 2.11$ and $n_a = 1.62$ determined previously by spectroscopic ellipsometry (SE) in the near IR range are employed. The a and b directions defined via the minimum and maximum intensity values in Figure 4.10 almost coincide with the directions of the electric field vector during measurement. They correspond to the [011] and $[0\bar{1}1]$ directions of the substrate, respectively.

Furthermore, we can estimate the ratios as

$$D_y = \frac{I_{by}}{I_{ay}} \quad \text{and} \quad D_z = \frac{I_{bz}}{I_{az}} \tag{4.14}$$

Under the assumptions that the long axes of the molecules are parallel to the sample surface and that the molecules have a good preferential orientation with their long axis parallel to the [011] direction of the sample, we can determine in a further approach an average tilt angle θ of the molecular plane with respect to the sample surface from the quotient of D_y and D_z:

$$\tan^2(1-\theta) = \frac{D_y}{D_z} \tag{4.15}$$

With the quantities from Table 4.2, a tilt angle of $\theta = 62° \pm 6°$ is calculated. The angle is larger compared to that from NEXAFS spectroscopy. Sources of error are the low-intensity of IR features, the angular spread of IR light, and the assumption that the x-axis of molecules lies parallel to the sample surface.

The angle χ determined from Eq. (4.1) should be the same for all x modes. However, with Eq. (4.1) different angles χ are calculated from the different dichroic ratios D_x of the ring band at 1596 cm^{-1} and the side-group band at 1692 cm^{-1}. The deviation is too high to be explainable by the experimental error alone. Further investigations are required for a better understanding.

4.4 Ellipsometry

From ellipsometric measurements, the components of the refractive index and extinction coefficient in different directions with respect to the substrate and their dependence on energy, the film thickness, and surface roughness are determined. The measured ellipsometric data are expressed in terms of the effective dielectric function

$$\langle\varepsilon\rangle = \sin^2\Phi + \sin^2\Phi\tan^2\Phi[(1-\rho)/(1+\rho)]^2 \tag{4.16}$$

where $\rho = r_p/r_s = \tan(\Psi)e^{i\Delta}$ is the complex reflectance ratio [190], Φ is the angle of incidence of light, Ψ and Δ are the so-called ellipsometric angles, and r_p and r_s are the Fresnel coefficients of the light-polarized parallel and perpendicular to the plane of incidence, respectively. The effective dielectric function is influenced by the optical properties of the film, its thickness, the dielectric function of the substrate, and the angle of incidence.

4.4.1 Optical Constants of PTCDA and DiMe-PTCDI

For the extraction of the complex refractive index components, the ellipsometry data were evaluated describing each sample by a layer model: substrate/optically anisotropic thin film/surface roughness. For the substrates, tabulated database values for the optical constants were used [191]. For PTCDA on GaAs, the anisotropy in the plane parallel to the substrate is found to be very small [156]. The imaginary parts of the effective dielectric function for the [011] and $[0\bar{1}1]$ directions differ only slightly. Consequently, the thin PTCDA films are modeled as being optically uniaxial and isotropic in the substrate plane.

In addition, film thickness and anisotropic optical constants are determined using a multiple-sample analysis procedure [162]. For this purpose, sets of three PTCDA layers with different thickness were prepared under the same growth conditions on GaAs and on Si substrates. During the fit procedure, the VASE data of all three samples on the same kind of substrate are calculated simultaneously with the optical constants of all the layers being coupled and so varied in the same way.

In contrast to PTCDA, for DiMe-PTCDI layers on GaAs, a strong in-plane anisotropy has to be considered additionally. Therefore, a model capable to describe an optically biaxial layer was applied. For the calculations, three data sets measured in anisotropic mode at different azimuthal orientation of the same DiMe-PTCDI sample were coupled and fitted together [169].

In a first step, the transparent spectral range was chosen for the determination of film thickness and surface roughness. Simultaneously, the thickness and values for the refractive index components with respect to the Cartesian axes in the substrate plane and normal to it are calculated. In the absorption-free low-energy range, the wavelength dependence of each refractive index was described by a Cauchy dispersion formula:

$$n = A_n + \frac{B_n}{\lambda^2} + \frac{C_n}{\lambda^4} \tag{4.17}$$

where A_n, B_n, and C_n are the Cauchy parameters and λ is the wavelength of light.

In addition to film thickness and surface roughness as well as Cauchy parameters for each refractive index, the procedure yields the orientation of optical axes for the biaxial DiMe-PTCDI film.

The next step of analysis is the determination of complex optical constants in the remaining spectral range. Therefore, thickness and surface roughness of the films were fixed and starting from the lower energy side, a point-to-point fit was carried out resulting in parameter sets of optical constants in each direction. In a last step, a model was built describing the optical constants of the DiMe-PTCDI film by a set of Gaussian functions, again keeping thickness and surface roughness constant.

Figure 4.22 shows the complex refractive indices for PTCDA parallel (ip) and perpendicular (oop) to the substrate plane as a function of energy. For comparison, spectra for some PTCDA films deposited on hydrogen-passivated Si(111) are shown. The hydrogen passivation has been achieved by etching the substrates for 2 min in HF(40%).Comparing the data sets for PTCDA on GaAs and Si, respectively, the line shapes are similar, but lower values for the refractive index for PTCDA on GaAs, especially for the in-plane component, are evident. This indicates a difference in microstructure and density with the data for PTCDA on Si being more characteristic for "bulk" PTCDA. The lower values of the refractive index for PTCDA on GaAs are in agreement with the island-like structure observed for thinner films. As for the refractive index, the extinction coefficient for PTCDA on Si is larger than for PTCDA on GaAs. Furthermore, the in-plane extinction coefficients are much larger than the out-of-plane ones.

The extinction coefficient k in the substrate plane shows a double feature with a smaller sharp peak at 2.25 eV, a broader main peak at 2.63 eV, and a weaker feature at 3.38 eV. These features are assigned to the transitions: S_0–S_1 (HOMO–

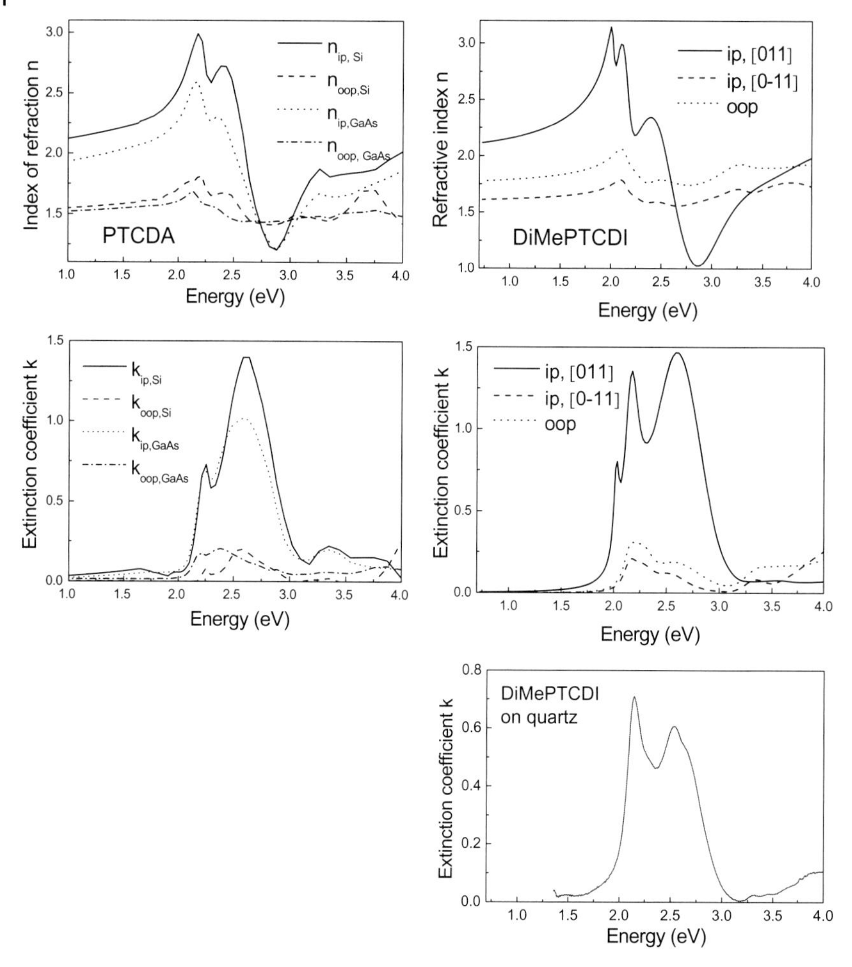

Figure 4.22 The optical constants of PTCDA films on Se-passivated GaAs(100) and on H–Si(111) (left side) and of DiMe-PTCDI films on S-passivated GaAs (100) (right side). For comparison, the extinction coefficient of an isotropic layer is presented. The notations ip and oop stand for in-plane and out-of-plane, respectively.

LUMO), S_0–S_1 with vibronic progression [165], and S_0–S_2 (the next highest dipole allowed transition) [164]. The optical constants obtained from a 120-nm-thick DiMe-PTCDI film on S-passivated GaAs(100) are likewise displayed in Figure 4.22. For the refractive index, the highest anisotropy is detected for the two distinct directions in the sample plane. The out-of-plane index value in the low-energy range lies in between.

The larger extinction coefficient k in the substrate plane shows a sharp double feature at 2.19 eV, and a broader main peak at 2.6 eV and weak features above 3 eV. These features are assigned to the transitions: S_0–S_1 (HOMO–LUMO) and S_0–S_1 with vibronic progression in analogy with PTCDA. The extinction coefficient obtained from a 50-nm-thick isotropic film on quartz is likewise displayed in Figure 4.22. This sample, which is isotropic due to a random orientation of molecules, exhibits only two peaks. This might be a hint that the first peak of the anisotropic sample at 2 eV is an artifact caused by interference. However, further measurements using samples with different thickness are required to clarify the origin of this feature.

4.4.2 Optical Constants of Crystalline and Thin-Film Pentacene

Prior to thin film growth, the pentacene was purified by twice by gradient sublimation. Then thin films were prepared in UHV on Si wafers with a thermally grown oxide of 200 nm thickness [192]. The twice sublimated source material was also used to grow pentacene crystals by the gas-flow reactor method [193].

X-ray studies performed in the Bragg–Brentano geometry show that as a function of the substrate temperature, different polymorphs can be distinguished. At reduced substrate temperatures and low coverage, the so-called thin-film phase characterized by a (001) lattice spacing of 15.5 Å is found. At higher substrate temperatures additionally, the so-called bulk phase starts to evolve with a (001) spacing of 14.4 Å. The single-crystal phase, on the other hand, has a (001) spacing of only 14.1 Å.

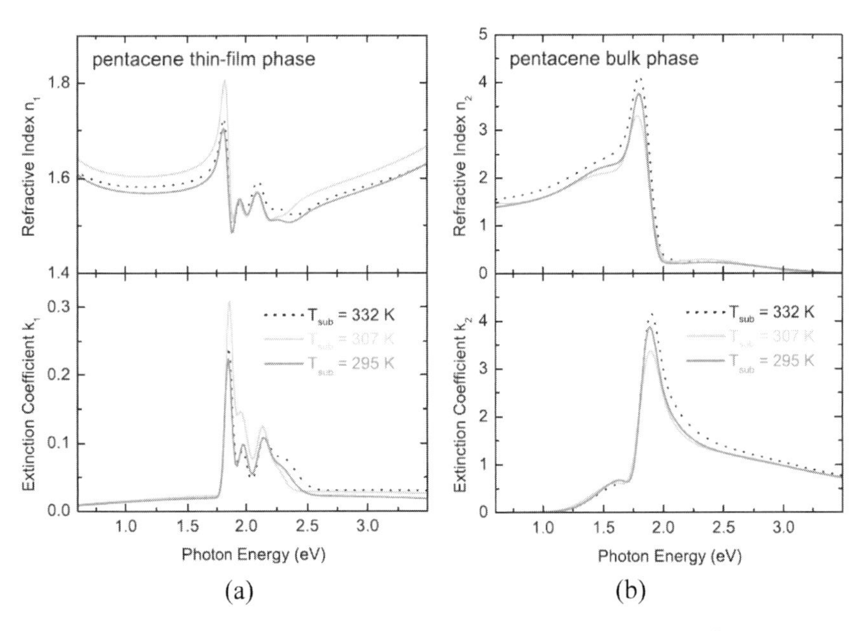

Figure 4.23 Real and imaginary parts of the refractive index, n_1/k_1 and n_2/k_2, of the pentacene thin-film phase and bulk phase, respectively. The optical properties vary only slightly with the substrate temperature.

The ellipsometric thin film spectra could be modeled by assuming optical isotropy for each layer and the following layer structure: Si-wafer/SiO_2/smooth pentacene layer 1 with the optical constants n_1 and k_1/rough pentacene layer 2 with the optical constants n_2 and k_2. In this model, layer 1 is described by five Gaussian oscillators and layer 2 by an effective medium layer with 27% voids and three Gaussian oscillators. For none of the pentacene films, prepared at different sample temperatures, a conversion from p to s-polarization is found; hence the ellipsometric data explicitly exclude an in-plane anisotropy.

The n and k values obtained in this way for films prepared at different sample temperatures are plotted in Figure 4.23 as a function of frequency. Layer 1 is identified as the thin film phase and layer 2 as the bulk phase. The optical response of the two polymorphs is clearly different, whereas the differences between several substrate temperatures within one polymorph are very small. Each crystallographic phase has its distinct optical signature independent of the thickness.

The data for the pentacene crystal was fitted using two Gaussian oscillators for the electric field vector $E\|y$ and $E\|z$ and three were used for the polarization $E\|x$. The respective data are shown in Figure 4.24.

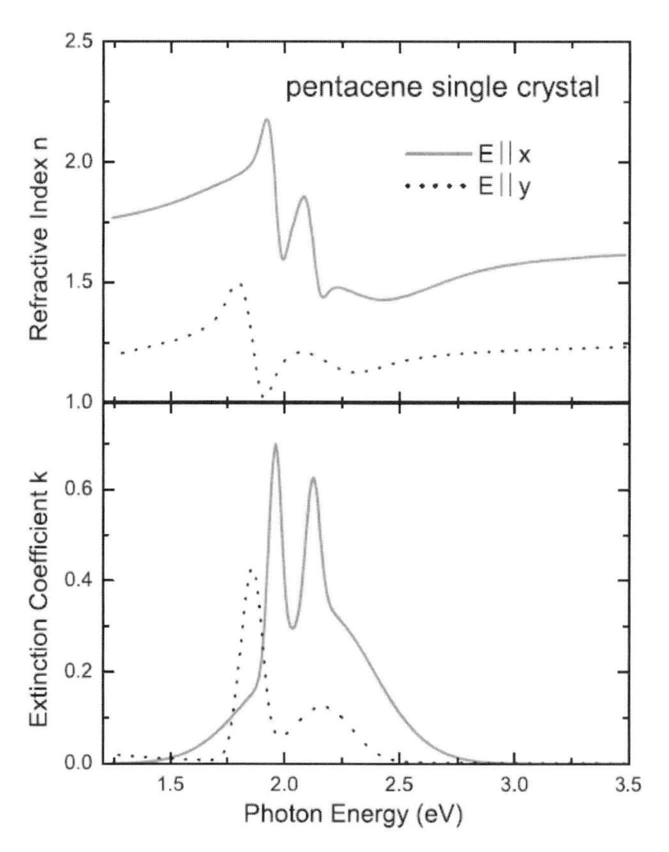

Figure 4.24 Optical constants of pentacene single crystals parallel to the x and y principal axes as obtained by generalized spectroscopic ellipsometry.

4.5 Reflection Anisotropy Spectroscopy

For the reflection anisotropy spectroscopy (RAS) measurements, the spectrometer used is similar in design as that described by Aspnes [194]. The spectrometer is operating at near normal incidence in the photon energy range from 1.8 to 5.3 eV. In RAS, the value $\Delta r/r = 2(r_\alpha - r_\beta)/(r_\alpha + r_\beta)$, which is the relative difference in the complex reflectance, r_α and r_β, for the light polarized along α and β directions perpendicular to each other in the surface plane, is measured. Here, the two directions α and β correspond to the $[\bar{1}10]$ and [110] directions in the surface plane of GaAs(001), respectively. As a result, RAS predominantly probes the optical anisotropy in the plane parallel to the surface. For cubic materials, the bulk is optically isotropic and the signal observed originates at surfaces and interfaces, which have reduced symmetry. For materials of lower (than cubic) symmetry as the organic compound PTCDA, the bulk can also contribute, since the symmetry in the plane parallel to the surface is lowered. As discussed previously, one also has to take into account interference effects, which can become quite large under certain conditions and in spectral regions where the films are transparent [195].

A typical reflectance anisotropy (RA) spectrum measured for a sulfur-terminated GaAs(001) surface is shown in Figure 4.25 and represents an RA fingerprint for the sulfur-induced (2×1) reconstruction of the GaAs surface. The sharp derivative-like feature around 3 eV is attributed to the E_1 and $E_1 + \Delta_1$ critical points of GaAs [196], while the broad dip at 4.7 eV is associated with the E_0' and E_2 critical points of GaAs. The feature near 3.5 eV is related to the sulfur passivation. Its origin is, however, still under discussion. While Hughes *et al.* attributed its occurrence to a transition involving sulfur dimers [197], other explanations like transitions within Ga–S bonds are also possible [198].

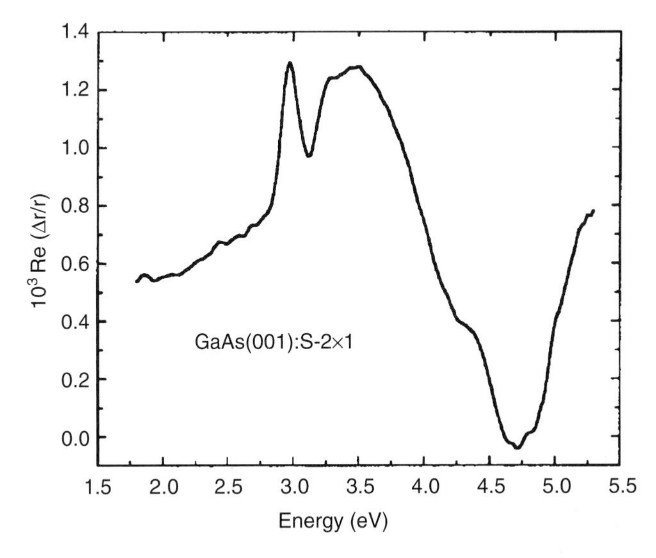

Figure 4.25 RAS spectrum of sulfur-passivated GaAs(001).

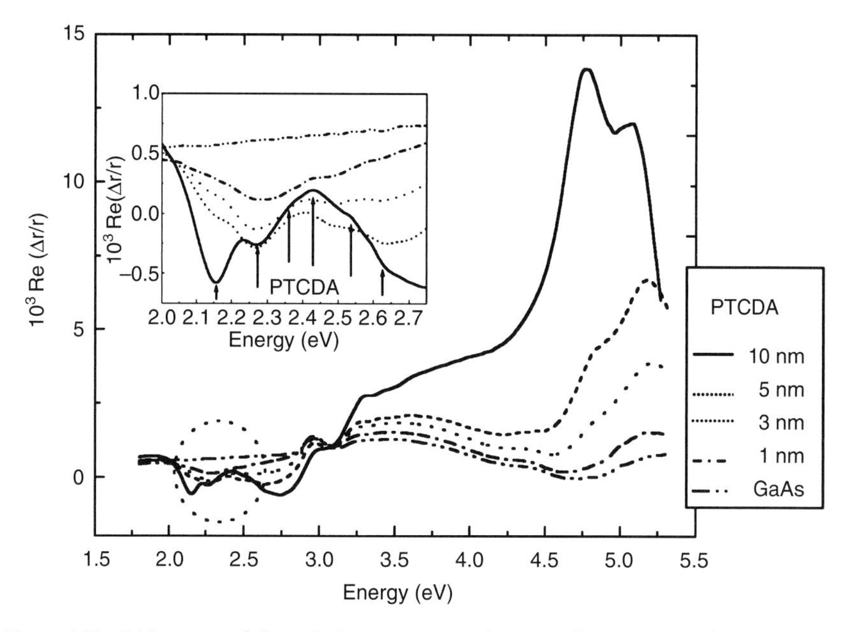

Figure 4.26 RAS spectra of clean chalcogen-passivated GaAs and upon PTCDA deposition.

The evolution of RA spectra with increasing PTCDA coverage is demonstrated in Figure 4.26 for layers with nominal thickness of 1, 3, 5, and 10 nm. Note that the features in the RA spectrum of clean GaAs appear as very weak in this figure, clearly illustrating the pronounced optical anisotropy induced by the presence of the PTCDA layer. Features indicated by arrows can clearly be assigned to PTCDA and their energy positions are listed in Table 4.3. It is important to note that these positions do not necessarily coincide with transition energies observed in optical absorption or ellipsometry (Im⟨ε⟩) spectra due to the low symmetry of PTCDA. Still a comparison of the region enlarged in the inset of Figure 4.26 with optical absorption spectra reveals good correspondence corroborating the assignment to PTCDA. The observation of a finite RAS response in the energy range 2–2.7 eV where no GaAs-related features are present and the anisotropy of the starting surface is small rules out that crystallites are completely random in their azimuthal orientation. The RA line shape of the GaAs-related feature at 3 eV hardly changes by deposition of PTCDA, thus suggesting that the underlying GaAs surface is quite likely to remain intact, and chemical interaction between organic molecules and the sulfur-passivated substrate is weak.

PTCDA features discussed so far are weak compared to the RA features observed above 4 eV, which show a dramatic increase with coverage. It was already shown that interference effects are predominantly contributing to these high–energy-range features [195]. Unfortunately, optical absorption data for PTCDA are not available for this energy range.

5
Electronic and Chemical Surface Properties

The experimental techniques discussed here are used to investigate the electronic and chemical properties of surfaces and interfaces. The electronic properties are, for example, of importance for the formation of injection barriers for charge carriers.

Figure 5.1 shows the energy level alignment in a GaAs/PTCDA/Ag heterostructure. Here, GaAs(001) serves as a substrate for the deposition of a thin PTCDA film. The GaAs is n-type doped and the surface will show a band-bending ($e_0 V_s$) region at the surface, which is depleted of majority charge carriers, that is, electrons. This band bending is due to acceptor-type surface states. After the formation of the GaAs/PTCDA, the energy level alignment at this interface will determine the efficiency of charge carrier injection into and from the PTCDA layer. It is of particular interest, if the band bending in the GaAs layer is changed upon PTCDA deposition and how the HOMO and LUMO of the PTCDA align with the valence band maximum (VBM) and conduction band minimum (CBM) of the GaAs. Preparing a metal contact on the PTCDA, the distance of the Fermi level with respect to the HOMO and LUMO will determine the injection barriers for electrons and holes.

Another point of interest is the existence of any band-bending region in the organic film. It should be mentioned that the injection currents can scale exponentially with the height of the interface barrier and changes in the barriers by a few 10 meV change current densities by orders of magnitude. Therefore, precise knowledge on the energy level alignment at organic devices is essential for the design of organic devices.

Quite often, the vacuum level alignment rule is used to determine the energy level alignment at organic interfaces. Here, it is assumed that the vacuum levels of the materials in contact align at the interface and the interface barrier heights can simply be calculated using the ionization potentials and electron affinities of semiconducting materials and workfunctions of metallic materials. For metal–organic interfaces, a conclusion that in general the vacuum levels do not align has been reached [199, 200]. The difference in vacuum levels is attributed to interface dipoles, and using photoemission spectroscopy (PES) values between 0.5 and 1 eV have been found for several metal/organic interfaces. One has to distinguish between interfaces obtained by growing organic molecules on metal substrates

Low Molecular Weight Organic Semiconductors. Thorsten U. Kampen

ISBN: 978-3-527-40653-1

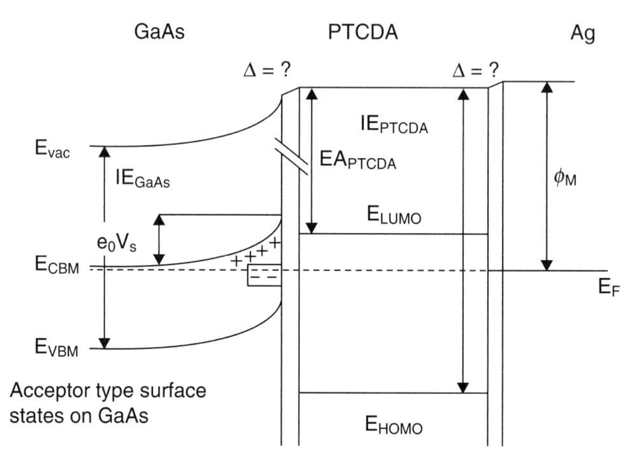

Figure 5.1 Energy level diagram of a GaAs/PTCDA/Ag heterostructure. IE and EA are the ionization energy and electron affinity, respectively. Δ denotes an interface dipole, while ϕ is the workfunction of the metal.

and metals grown on organic thin films. The latter case may show different energy level alignment because single metal atoms or small clusters impinging on organic surfaces can possess chemical properties different from those of a solid metal surface. In addition, metal atoms can diffuse into the organic films.

Ishii *et al.* have discussed the origin of interface dipoles at organic-on-metal interfaces in detail [201]. The mechanisms they proposed are briefly presented in a sequence of increasing interaction. In the case of physisorption, only a weak interaction takes place, which results in redistribution of charge in the adsorbate and/or substrate surface. These mechanisms are always operative, but may be enhanced or compensated by chemisorption effects. In the case of chemisorption, a charge transfer between substrate and absorbate takes place. This charge transfer can be due to the formation of charge transfer complexes or the formation of a covalent/ionic bond between the adsorbed molecule and the substrate surface. The former and latter case will be called intermediate and strong interaction, respectively.

Weak interaction – physisorption

In this regime, the interaction between the substrate and the molecule is restricted to van der Waals interaction. Even though van der Waals interaction is often considered to be very small, it leads to the formation of molecular crystals and changes in electronic surface properties. A well-known example for the latter case is the adsorption of rare gases on metal surfaces. The physisorption of Xe on different clean metal surfaces induces a decrease in the metal workfunction [202]. These changes in the metal workfunctions correspond to an interface dipole at the Xe/metal interface, which has been found to increase with increasing metal workfunction.

These workfunction changes can be explained in two different models, that is, adsorption-induced dipole in the adsorbate [203] and the push-back effect. The adsorption-induced dipole is due to the motions of electrons in the Xe atom. This motion results in an alternating dipole moment, its field being screened by the metal electrons resulting in an image force potential. As a result, the Xe electrons are shifted toward the metal. Therefore, the adsorbed Xe atoms exhibit a permanent dipole moment with its positive side pointing toward the vacuum. This effect is independent of the metal and decreases the workfunction of the metal surface. Calculations of the charge density distribution in a local density approximations (LDA) for Xe adsorbed on a metal predict such displacements of electrons [204, 205]. The push-back effect, which is also called pillow effect, is due to the fact that the electron density does not drop abruptly to zero at a metal surface. Within a metal, described in a jellium model, the electron density is constant. At the surface, the electron density shows Friedel oscillations on the metal side and an exponential decay into the vacuum. The resulting dipole layer at the metal–vacuum interface is negative on the vacuum side. Due to the adsorption of the Xe atoms, this electron density is pushed-back into the metal decreasing the potential drop at the surface. This is equivalent with the formation of an Xe-induced interface dipole, having its positive side directed toward the vacuum and reducing the metal workfunction.

An example for weak interaction is the TTC/metal interfaces. TTC (n-$C_{44}H_{90}$) is a long alkane chain with a wide band gap of about 9 eV [206]. The ionization energy of 8.5 eV [207] results in the LUMO being positioned above the vacuum level. TTC is found to physisorb on metals like Au, Ag, Pb, and Cu with the vacuum level on the TTC side of the interface being lower than that on the metal side [208–210].

An additional effect, which has to be taken into account, are polar organic molecules having a permanent dipole moment. This dipole moment will contribute to a surface dipole if it has a component perpendicular to the surface.

Intermediate interaction – chemisorption and formation of charge transfer complexes

Charge transfer between an adsorbate and a substrate surface can result in interface dipoles of both signs. For the interface between a strong acceptor molecule and a low workfunction metal, the formation of an anion is expected, while a cation is created for the adsorption of a donor molecule on a high workfunction metal. Therefore, anion and cation formation causes an increase and decrease in vacuum level, respectively.

One mechanism resulting in a charge transfer across organic/metal interfaces is the dynamical charge transfer (DCT). The DCT requires a vibration that modulates the energy of a molecular orbital, which is partly filled and hybridized with the substrate electronic wavefunctions. Here, the vibration of the molecule induces a charge oscillation between the substrate and the molecule. The generated oscillating dipole is oriented perpendicular to the surface and can couple to an electromagnetic wave.

Strong interaction – chemisorption and formation of covalent/ionic bonds

Here, the chemisorption results in the formation of chemical bonds between the molecule and the substrate having a covalent or ionic character. The ionicity of the bond depends on the difference in electronegativity of the molecule and the substrate. Strong interaction occurs on surfaces of sp-metals, d-metals on the left-hand side of the periodic table of elements, and elemental semiconductors. It is more likely to occur for low-symmetry molecules, heterogeneous molecules, molecules with functional side groups or unsaturated bonds, and molecules with high ionicity. For example, PTCDA partly dissociates and adsorbs with random orientation on Si(100) and Ge(100) [211]. Another example is the adsorption of organic amines on elemental semiconductor surfaces. On Ge(100)-(2×1) surfaces, nonaromatic amines form a stable dative-bonded species at room temperature [212, 213]. Dative bonding, also known as coordinate covalent bonding, occurs when one molecule donates both of the electrons needed to form a covalent bond.

Another important issue is the occurrence of a "band-bending"-like electrostatic energy shift in organic layers, which has been observed in many metal/organic systems [214]. In most cases, this shift is confined to a regime of only a few nanometers, which cannot be accounted for using the conventional band-bending theory of inorganic semiconductors. Shifts occurring in such small thickness ranges can be due to a change in the intermolecular interaction, namely, due to a change in the molecular orientation as a function of the film thickness.

In general, the spatial distribution of the electron wavefunctions is not isotropic, and therefore the overlap of substrate and molecule wavefunctions depends on the orientation of the molecule. For planar molecules like PTCDA with a π-electron system, a flat-lying orientation on a substrate results in a stronger interaction with the substrate than an upright orientation. The energy position of the HOMO depends on the charge transfer across the interface and the intermolecular interaction, because this will induce a change in the polarization. The change in the molecular orientation thus produces a band-bending or energy level shift that plays an important role in the actual device properties. This observation has been made for copper phthalocyanine (CuPc) films grown on MoS_2 substrates [215]. Angular resolved ultraviolet PES using synchrotron radiation has been used to determine the energy position of the HOMO and the orientation of the molecules as a function of the film thickness. For a layer thickness of 0.4 nm, the HOMO is found to be $\approx$1.05 eV above the Fermi level, while for films of 8.4 nm thickness the HOMO shifts by about 0.3 eV to higher binding energies. Following the intensity of the HOMO as a function of emission angle and azimuth, the tilt angle of the CuPc molecular plane with respect to the substrate surface is determined to be $0°$ and $10°$ for the 0.4 and 8.4 nm films, respectively.

In metal-on-organic systems, metal atoms diffusing into the organic film can also induce a band-bending-like effect. If the interaction of the metal with the organic film is of ionic nature, the metal ions form a gradient within the organic layer and an effect similar to band bending may be observed [216, 217]. Here, the evolution of core and valence spectra during the deposition of a vinylene phenylene oligomer on Ca has been studied using PES. An energy level bending of 0.5 eV in a thickness of about 10 nm was observed.

Another physical mechanism resulting in shifts of the Fermi level are gap states, which are the most relevant mechanisms at inorganic semiconductor interfaces. For organic materials, the existence of interface gap states is currently under investigation. They have been observed in long-chain alkane films [218] and at interfaces of Mg, Ca, or Li on Alq_3 [219–221], and Ag on 3,4,9,10-perylenetetracarboxylic bisimidazole (PTCBI) [222]. These states are often refereed to as polaron or bipolaron states. Negative polarons result from filling the former LUMO with one electron, leading to half-filled orbital. This results in new electronic states in the normally forbidden energy gap, having a finite density of states at the Fermi energy. In the case of bipolarons, the former LUMO is filled with two electrons, and the new occupied levels are expected to be below the Fermi level. Quantum chemical calculations show that the two new intragap states of the bipolaron are located deeper in the gap than the polaronic states [223].

Hirose *et al.* investigated the chemical and electronic properties of interfaces between PTCDA films (40–80 nm) and In, Al, Ti, Sn, Au, and Ag using synchrotron radiation PES [224]. In, Al, Ti, and Sn are found to react at room temperature with the oxygen containing anhydride group of the PTCDA molecule. The results of these reactions are interface states in the band gap of PTCDA. The penetration of the reactive metals is found to be inversely related to their first ionization energy, whereby ionized metals are driven into the PTCDA film by Coulomb repulsion. For In, the penetration depth is found to be above 10 nm. All reactive metals result in ohmic contact, which is attributed to carrier conduction through interface gap states. For the noble metals Au and Ag, no interface reaction and diffusion is observed which results in a blocking concerning carrier transport. The In/PTCDA system was investigated in more detail by Azuma *et al.* [225]. Here, only a monolayer of PTCDA was prepared on cleaved MoS_2. They also observed In-induced interface states in the band gap of PTCDA. These states are attributed to the reaction between In and PTCDA resulting in a In_4–PTCDA compound. A comparison of molecular orbital calculations and angular resolved photoemission measurements shows that the gap states can be explained by the calculated density of states of the In_4–PTCDA complex, where the gap states originate from the π-orbital which consists of carbon $2p_z$, oxygen $2p_z$, and indium $5p_z$ atomic orbitals. Cs deposition on PTCDA is also known to produce gap states, observed in the VB spectra by Ertl *et al.* [226]. However, their MO calculation shows that even without a covalent bond between Cs and PTCDA, the charge transfer can occur through O atoms in the anhydride and the carboxylic group. As such, it is still under debate whether and where in the molecules a chemical reaction occurs at metal/PTCDA interfaces.

Even a continuum of interface states, which are observed at inorganic semiconductor surfaces and interfaces, are currently considered to be relevant for the electronic properties of metal/organic interfaces. Vázquez *et al.* have investigated the metal/PTCDA interface barrier theoretically using weak chemisorption theory [227]. They found an induced density of states even for a weak metal/PTCDA interaction. The induced density of states is found to be large enough for the definition of a charge neutrality level (CNL), which is located 2.45 eV above the HOMO. The charge transfer across the interface is found to be due to tunneling of metal electrons through the molecular gap.

It should be mentioned at this point that recent investigations have shown that even Fermi level alignment is not achieved for some organic/metal interfaces. The workfunction as a function of thickness of *N,N'*-bis(3-methylphenyl)-*N,N'*-diphenyl-[1,1'-biphenyl]-4,4'-diamine (TPD) films grown on Ca, Mg, Ag, Cu, and Au substrates was measured with a Kelvin probe [201]. For all interfaces, an abrupt decrease in workfunction was observed within the first 1 nm of film thickness. Further deposition up to 100 nm resulted in no changes in the workfunction. The final values reached differ significantly. Here, Ca and Au result in the lowest and highest workfunction of about 2 and 4 eV, respectively.

Also for organic/organic interfaces, the formation of interface dipoles has been found, but they are smaller than at organic–metal interfaces. According to Ishii *et al.* [228], interface dipoles are found at interfaces between organic materials with greatly differing ionization energies and electron affinities. Here, a charge transfer is expected from the low ionization energy (donor) molecule to the high electron affinity (acceptor) molecule. This argument is also at the basis of the work of molecular doping [229]. This argument does not explain the interface dipoles for all organic/organic interfaces. Experimental investigations give an interface dipole of only 0.1 eV for the PTCDA/α-NPD interface, while, according to the argument given above, one would expect a larger interface dipole because of the ionization energy and electron affinity of *N,N'*-diphenyl-*N,N'*-bis(1-naphthyl)-1,1'-biphenyl-4,4''-diamine (α-NPD) being more than 1 eV smaller than the respective values of PTCDA. Such deviations may be attributed to high densities of impurity or defect-induced states in the band gap of the organic materials.

For organic/inorganic semiconductor interfaces, however, no systematic study was performed on the energy level alignment and all the questions mentioned above remain unsolved. The energy level alignment for interfaces of PTCDA on GaAs(100) and CuPc and α-NPD on InP(110) was reported by Hirose *et al.* and Chasse *et al.* [230], respectively. In all cases, the vacuum level alignment rule does not hold and interface dipoles range between +0.6 eV for GaAs/PTCDA and −0.78 eV for InP/CuPc.

5.1
Photoemission Spectroscopy

In PES, electrons are excited from occupied states below the Fermi level into unoccupied states above the vacuum by light in UV and X-ray regime. As shown in Figure 5.2 using a monochromatic light source, the binding energy of electrons in the sample can be determined from the distribution in kinetic energy above the vacuum level. The excitation of the electron follows energy as well as momentum conservation:

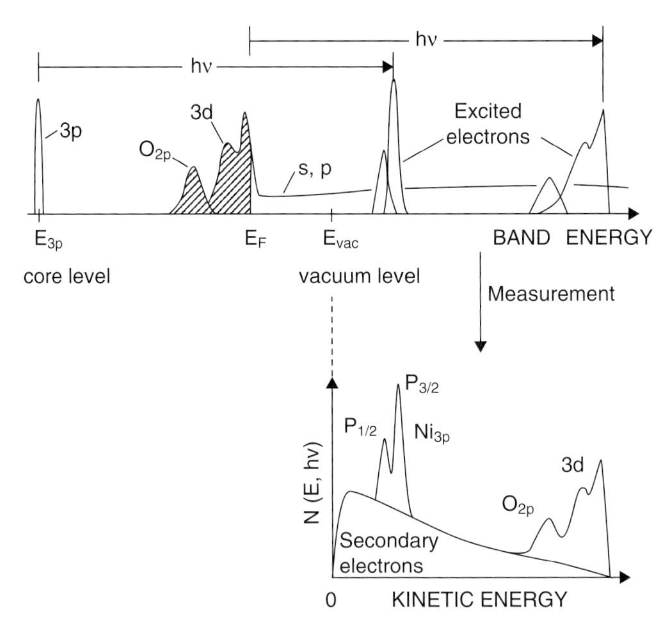

Figure 5.2 Principle of photoemission spectroscopy.

$$E_{\text{kin}} = \hbar\omega - E_i - \Phi \tag{5.1}$$

$$\mathbf{k}_f = \mathbf{k}_i + \mathbf{q}_{\text{ph}} \approx \mathbf{k}_i \tag{5.2}$$

In Eq. (5.1), E_{kin} is the kinetic energy of electrons in the vacuum, $\hbar\omega$ is the photon energy, E_i the binding energy of electrons in the solid, and Φ the workfunction of the sample. The wavevector of the photon $\mathbf{q}_{\text{ph}}$ is negligibly small compared to the wavevector of the electron in initial ($\mathbf{k}_i$) and final states $\mathbf{k}_f$.

The emission from core levels at higher binding energies may be used to investigate the chemical composition of the sample, and its chemical interaction with the environment. A charge transfer accompanies usually chemical bonding between the atoms or molecules involved in the bonding. Here, charge is transferred from the less electronegative to the more electronegative atom/molecule. The change in charge will change the binding energy of electrons, resulting in an energy shift of core levels in the photoemission spectra.

The low binding energy states in the valence band reflect the electronic structure of the solid. Here, macroscopic surface and interface properties like workfunction or ionization energy can be determined. As shown in Figure 5.3, the ionization energy of PTCDA is determined by measuring the width W of the photoemission spectrum, that is, the energy distance between the low energy cut-off of the spectrum and the HOMO. A negative bias applied to the sample will shift the photoemission spectrum to higher kinetic energies, which will help to determine the low energy cut-off properly. The energy position of the HOMO is determined by

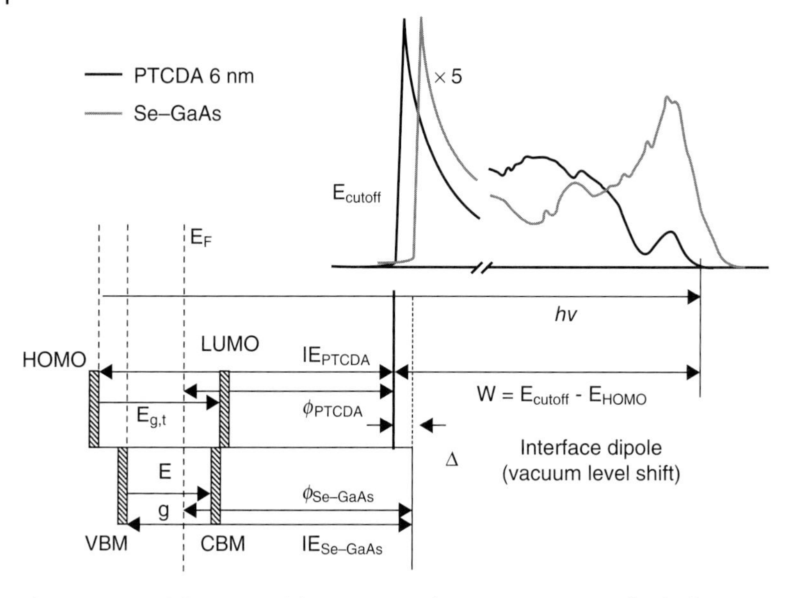

Figure 5.3 Band diagram and the respective photoemission spectra for the formation of a PTCDA/GaAs interface.

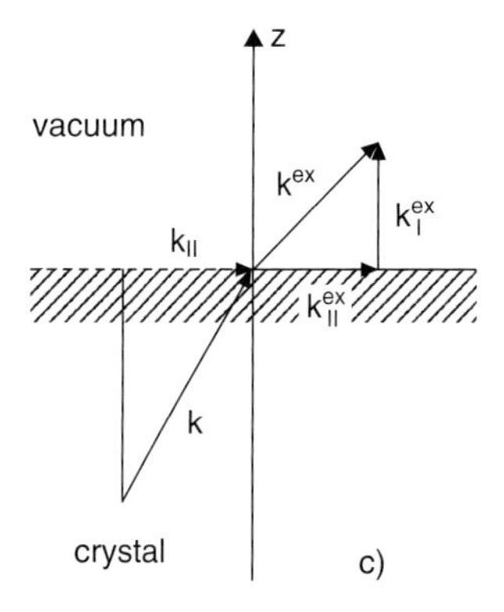

Figure 5.4 Change in wavevector of a photoelectron leaving a solid.

the extrapolation of the high-energy edge to zero intensity. This is in analogy to the determination of the valence band edge in inorganic semiconductors. Subtracting the width of the spectrum from the photon energy finally gives the ionization energy.

Performing angular dependent and photon energy dependent measurements, any dispersion in the valence states can be detected. The wavevector of the excited electron may be described by two components being parallel and perpendicular to the surface. When the electron leaves the solid, it has to overcome the workfunction, which results in a change of the wavevector component perpendicular to the surface (see Figure 5.4). The parallel component is preserved, on the other hand.

In vacuum, both components of the wavevector relate to the kinetic energy as

$$k_f^{\perp} = \sqrt{\frac{2m}{\hbar^2} E_{\text{kin}}} \cos\Theta, \quad k_f^{\parallel} = \sqrt{\frac{2m}{\hbar^2} E_{\text{kin}}} \sin\Theta \tag{5.3}$$

where m is the mass of an electron and Θ the detection angle with respect to the surface normal.

5.1.1 UPS on PTCDA and DiMe-PTCDA

Due to the localized electronic structure in organic semiconductor, their electronic properties are much easier to calculate than those of inorganic semiconductors. In the following, the partial density of occupied states (PDOOS) and the respective total density of occupied states (TDOOS) of PTCDA and DiMe-PTCDI have been calculated using Gaussian 98 B3LYP method and the 6-31++G(d,p) basis set [166] together with the AOMix software [231, 232]. Under the assumption that the electronic properties are predominantly described by the properties of a single molecule, the calculated molecular energy levels are compared with the experimentally determined densities of occupied states. The calculated molecular orbitals are broadened by a Gaussian function with 0.5 eV FWHM. The comparison of experimental and theoretical results for PTCDA and DiMe-PTCDI is shown in Figures 5.5 and 5.6.

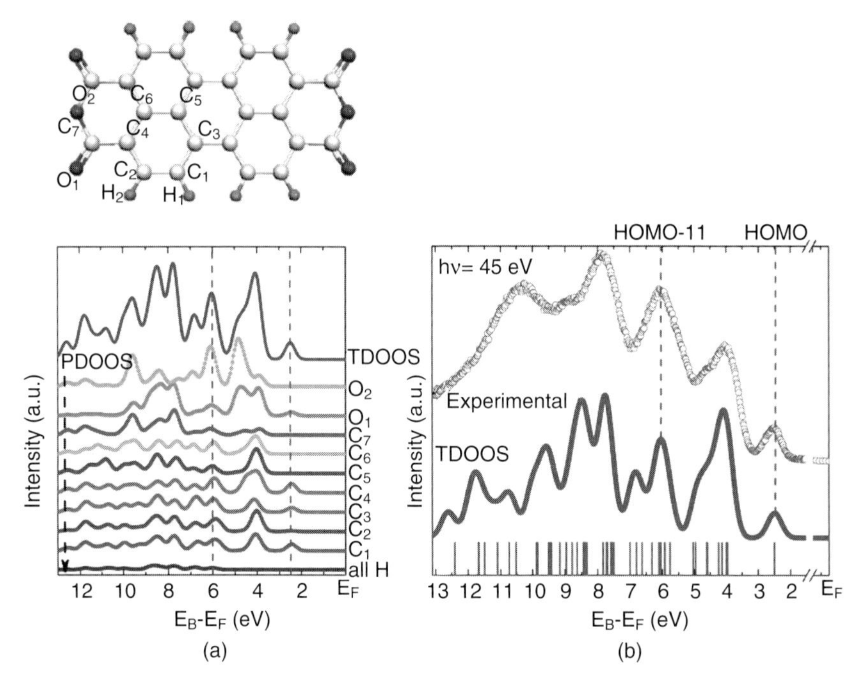

Figure 5.5 (a) PDOOS and TDOOS of PTCDA. (b) Valence band spectrum of a PTCDA thin film (symbols) along with the calculated DOOS (line). The MO energies are shown by vertical bars. Above: left, geometry of PTCDA with labels on the atoms for which the PDOOS is shown in (a). [from Reference 232]

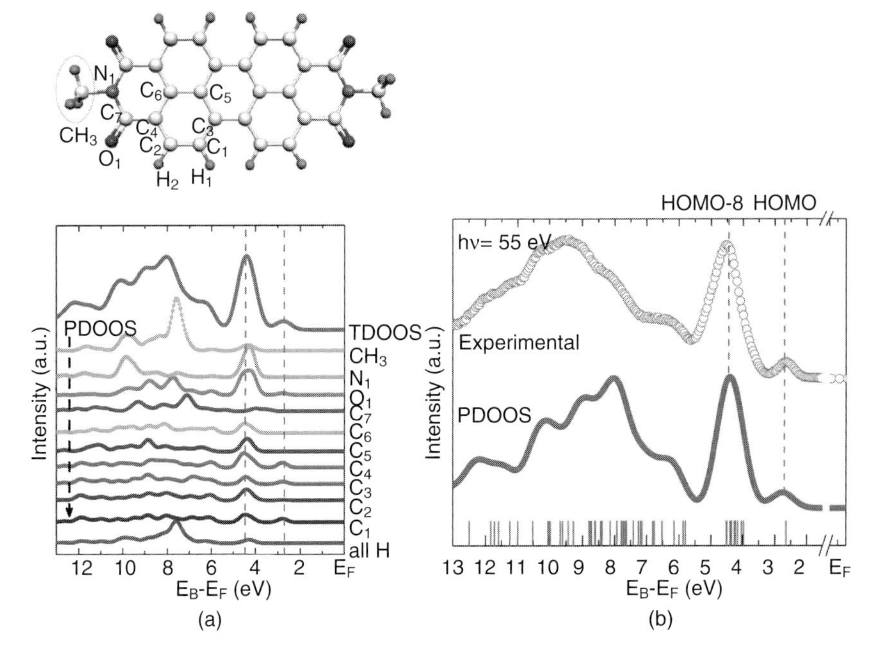

Figure 5.6 (a) PDOOS and TDOOS of DiMe-PTCDI. (b) Valence band spectrum of a DiMe-PTCDI thin film measured at 55 eV excitation energy and calculated TDOOS. The MO energies are shown by vertical bars. Top figure: left, geometry of DiMe-PTCDI with the labeled atoms that are contributing to TDOOS. [from Reference 232]

For PTCDA, the HOMO is π-orbital well separated from the other occupied molecular orbitals and mainly located at the perylene core. The next four bands are related to contributions with π character from the perylene core and to the molecular orbitals derived from oxygen $2p_x$ and $2p_y$ atomic orbitals. Moreover, the feature at 6 eV (HOMO-11) is mainly assigned to the contribution of the $2p_z$ atomic orbitals of the anhydride groups. The TDOOS derived from the PDOOS is compared to UPS spectra taken from PTCDA with an excitation energy of 45 eV. There is a good agreement between the experimental and theoretical data. It should be mentioned that the theoretical data have been shifted by 0.65 eV toward higher binding energy.

Looking at the PDOOS of DiMe-PTCDI, one notices immediately the dominant feature at about 4.5 eV. This feature is a consequence of the imide group and has contribution from both the methylimide group as well as from the carboxylic groups. The HOMO, as in PTCDA, is predominantly located at the perylene core of the molecule. The pronounced feature at 4.5 eV binding energy is also visible in the TDOOS, the latter one being shifted by 0.73 eV toward higher binding energy. Again, there is a good agreement between the experimental and theoretical data.

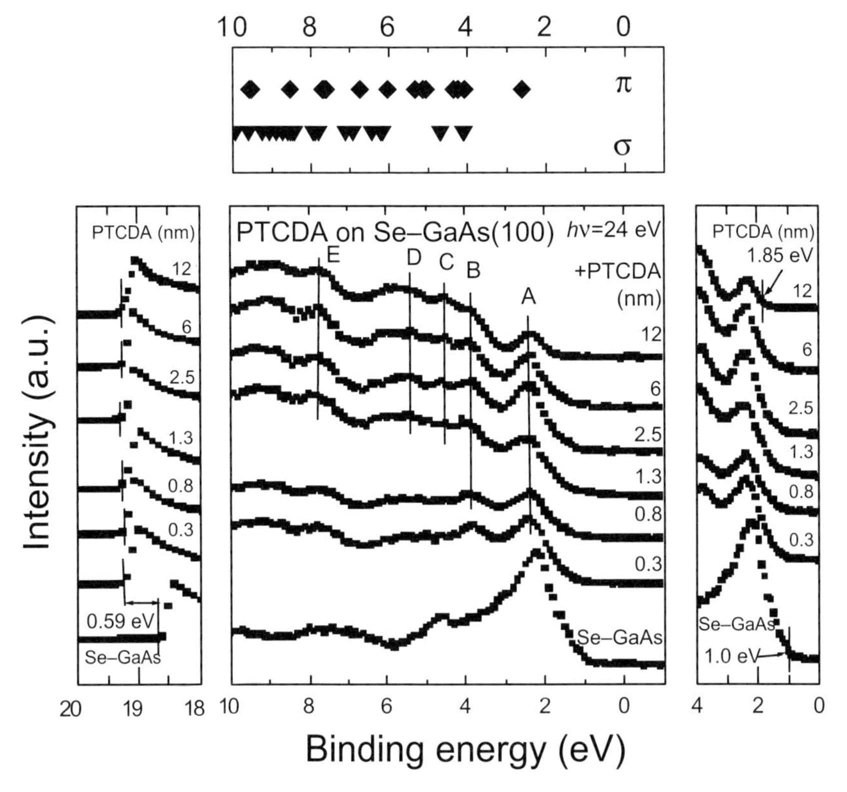

Figure 5.7 Valence band spectra of PTCDA deposited on Se-passivated GaAs for several thicknesses. Upper panel shows calculated energy positions of occupied π- and σ-orbitals. The spectra were measured with synchrotron radiation of $h\nu = 24\,\text{eV}$.

5.1.2
Energy Level Alignment at PTCDA and DiMe–PTDI Interfaces

Figure 5.7 shows UPS for the clean Se-passivated GaAs(100) surface and after subsequent stepwise deposition of a 12-nm PTCDA film onto this surface. In the spectra of a thick PTCDA films, five features labeled as A, B, C, D, and E are clearly seen. This agrees perfectly with previous work [225, 233]. For the assignment of the spectral features, the experimental data are compared to results from molecular orbital calculations using density functional theory. The calculations were performed using the Gaussian 98 package at the B3LYP level of theory with a standard 6-31G(d) basis set. The energy levels of all molecular orbitals in the respective energy ranges are shown in the upper panel. To align the highest occupied energy level with the center of the lowest binding energy feature in the UPS, all molecular orbitals were shifted by 0.65 eV to higher binding energies. Feature A with the lowest binding energy is attributed to the HOMO and originates from a single molecular orbital, which has a π-character and is distributed predominantly over the perylene core. The width of feature is quite large for a single molecular level. This broadening can be attributed to a spatial variation in intermolecular relaxation energy due to the presence of the surface and disorder. Molecular vibrations may also contribute to the broadening. The features B, C, and D have a predominant π-character, while predominantly σ-orbitals contribute to feature E.

The left and right panels show the onset of the high binding (low-kinetic) energy secondary electron peak and the valence band structures, respectively. The ionization energy of the substrate is obtained by subtracting the total width of the valence band spectra from the photon energy, that is, $IE_S = h\nu - (C_{cut\text{-}off} - E_{VBM})$. Here, $C_{cut\text{-}off}$ and E_{VBM} represent the energy positions of the secondary electron onset and the VBM of the substrate surfaces relative to E_F, respectively.

IE_{PTCDA} was obtained by replacing E_{VBM} by the low-binding energy edge of HOMO (E_{HOMO}) in the previous equation. The energy position of the HOMO as well as of all other PTCDA features do not change with increasing film thickness, indicating the absence of any band bending in the organic film. Therefore, shifts in $C_{cut\text{-}off}$ upon PTCDA deposition can be interpreted as interface dipoles.

The energy barrier for hole transport is given by the difference $E_{VBM} - E_{HOMO}$. The only energy positions that are not directly obtained from the measured UP spectra are E_{CBM} of the substrates and E_{LUMO} of the PTCDA films. In order to determine E_{CBM}, it is reasonable to use the optical gap of GaAs of 1.42 eV because of the low polarization energies and high carrier screening efficiency leading to exciton-binding energies of a few millielectron volts. This optical gap of GaAs(100) does not change upon different surface treatment since the modification is confined within a few atomic layers. On the other hand, E_{LUMO} of the PTCDA film is not yet well known. The difference between the optical and transport HOMO–LUMO gap of PTCDA films was investigated by Hill *et al.* using UPS and IPES. This value amounts to ~0.6 eV. The accuracy of this value is limited by a rather poor resolution of inverse PES (IPES).

Figure 5.8 shows the energy level alignment obtained from UP spectra between PTCDA films and GaAs(100) surfaces with different *IEs*. For simplicity, the band bending of the substrates is omitted so that the energy level positions corresponding to the substrates represent those of the substrate surfaces. It can be seen that the different surface treatment varies *IEs* of the GaAs(100) surfaces. The measured *IEs* ranges from (5.23 ± 0.10) eV for the GaAs(100)-c(4 × 4), (5.55 – 5.92 ± 0.10) eV for S–GaAs(100) to (6.40 ± 0.10) eV for the Se–GaAs(100) surfaces. The *IEs* of the GaAs(100)-c(4×4) surface agree well with the value of 5.29 eV reported previously [234]. It is known that the passivation of GaAs(100) surfaces by S or Se atoms terminates the chemically active sites, leading to the formation of S–Ga or Se–Ga surface dipoles with S or Se atoms on the surfaces, and thus inducing the change in IE_S. It should be noted that the scatter of IE_S observed for S–GaAs(100) surfaces can be correlated with the development of the VB structures. A slightly different temperature ramp rate and pressure increase during annealing appears to strongly affect the degree of the related surface reconstruction. On these surfaces, PTCDA

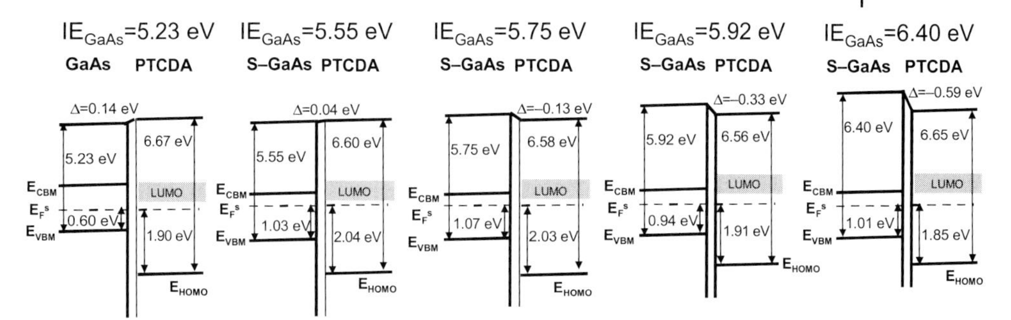

Figure 5.8 Energy level alignment at interfaces of PTCDA on GaAs(100) surfaces with different *IE*s: the shaded region represents the possible energy position range of the LUMO. The lower and upper limits are drawn considering the optical (2.2 eV) and transport HOMO–LUMO gap (2.8 eV) of the PTCDA. The transport gap was obtained by adding an energy difference of 0.6 eV between optical and transport gap.

was evaporated in a stepwise manner. The energy levels of the PTCDA films were obtained using the UP spectra of thin (4–12 nm) PTCDA films in order to prevent any influence due to sample charging.

The energy position corresponding to the center of the PTCDA HOMO in UP spectra for each sample does not appear to change during the stepwise deposition despite the fact that it is difficult to evaluate the energy shift below a PTCDA thickness of 1 nm due to the screening by the VB features of the substrates. This indicates that the PTCDA films show no "band-bending"-like behavior. The measured IE_{PTCDA} varies between 6.56 and 6.67 eV, and the Fermi level is found to be 1.85–2.04 eV above $E_{HOM.}$ However, taking into account the accuracy measurements (±0.1 eV) both properties are constant. Therefore, Fermi level alignment is achieved at the PTCDA/GaAs(100) interfaces investigated here. The shaded region in Figure 5.8 shows the possible energy position range for the PTCDA LUMO, with the lower limit being drawn using the optical HOMO–LUMO gap value of PTCDA (2.2 eV) obtained by the energy position of the first peak in the optical absorption spectra of PTCDA films. The upper limit is obtained considering an energy difference of 0.6 eV between the optical and transport HOMO–LUMO gap of PTCDA.

A strong correlation is found between the interface dipole and the relative energy position between E_{LUMO} and E_{CBM}, respectively, the EA of the PTCDA film (EA_{PTCDA}) and substrates (EA_S). E_{HOMO} is always located well below E_{VBM}. At the PTCDA/GaAs(100)-c(4×4) interface, where a positive interface dipole is formed, E_{LUMO} is located below E_{CBM}. The situation is reversed when a negative dipole is formed, as in the case of the PTCDA/Se–GaAs(100) interface. Consequently, the interface dipole formed at the PTCDA/S–GaAs(100) interface varies from positive to negative depending on IE_S (or EA_S). It can, therefore, be deduced that the formation of the interface dipole at PTCDA/GaAs(100) interfaces is possibly driven by the difference in EA_S and EA_{PTCDA} and that in general the vacuum level alignment rule

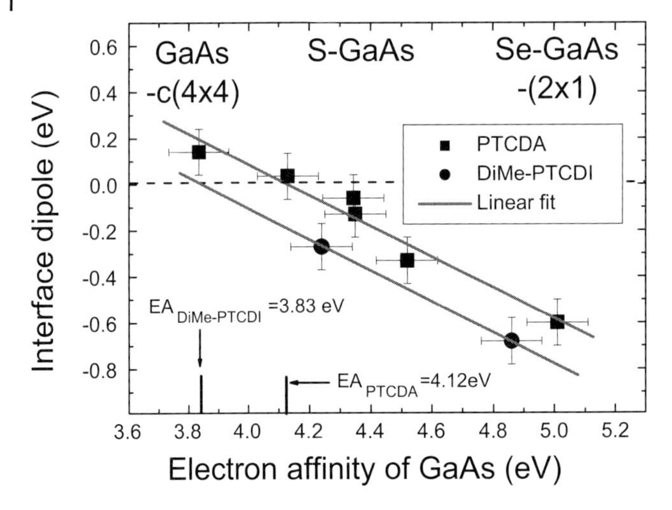

Figure 5.9 Interface dipole formed at PTCDA/GaAs(100) and DiMe-PTCDI/GaAs(100) interfaces versus electron affinity of GaAs(100) substrate surfaces.

is not applicable for those interfaces. At thermal equilibrium, the number of electrons and holes that are transported across the interfaces should be equal. Due to the difference in *EA* and *IE* between substrate surfaces and PTCDA films, each electron and hole transported undergoes an energy loss or gain. The net energy loss, therefore, depends on the electron and hole concentration that is transported across the interface and the energy difference of $EA_S - EA_{PTCDA}$ and $IE_S - IE_{PTCDA}$. The interface dipole is formed in order to compensate the net energy loss. In the case of PTCDA films on n-type GaAs(100) surfaces, it is expected that the number of electrons transported across the interface is much higher than that of holes and, therefore, $EA_S - EA_{PTCDA}$ can be proposed to be the determining driving force for the interface dipole formation.

In Figure 5.9, the interface dipole is presented as a function of EA_S. It can be seen that the interface dipole formed at PTCDA/GaAs(100) interfaces is linearly dependent on EA_S. Using a linear fit, the interface dipole is found to be zero at $EA_S = (4.12 \pm 0.10)$ eV. This value also represents EA_{PTCDA} assuming that the formation of the interface dipole is driven by the difference in EA_S and EA_{PTCDA}. Using $EA_{PTCDA} = (4.12 \pm 0.10)$ eV, the energy offset between E_{CBM} and E_{LUMO} at the interfaces can be estimated to be (-0.17 ± 0.10) eV for PTCDA/GaAs(100)-c(4×4), $(0.05 - 0.087 \pm 0.10)$ eV for PTCDA/S–GaAs(100), and (0.27 ± 0.10) eV for PTCDA/Se–GaAs(100) interfaces. In addition, assuming that the energy level of the PTCDA films extends up to the interfaces without energy shifts, we can estimate the HOMO–LUMO gap to be in the range of 2.44–2.55 eV. To compare this value with the one of 4 eV proposed by Hill *et al.*, one has to take into account that Hill *et al.* determined the transport gap from the energy distance of the maxima of HOMO and LUMO peaks recorded with PES and IPES, respectively. However, taking their spectra and determining the difference in leading edges of the HOMO and LUMO peaks, one obtains a value of approximately 2.3 eV, which is in good agreement with the value presented here.

In this study, the transport gap has been determined by analyzing the energy level alignment at PTCDA/GaAs(100) interfaces. Therefore, it will be used without any further modification for the interpretation of the IV characteristics of Ag/organic/GaAs(100) interfaces.

UPS spectra for the growth of DiMe-PTCDI on S-passivated GaAs(100) are shown in Figure 5.10. As for PTCDA, the HOMO consists of a single π-orbital. The major difference in the UP spectra from PTCDA and DiMe-PTCDI is the strong peak at a binding energy of ~4.5 eV in the spectrum of DiMe-PTCDI. This peak is the characteristic for DiMe-PTCDI and stems from the π- and σ-bonds located at the imide, carboxyl, and methyl groups of the molecule. For the clean substrate surface, the position of VBM in S-passivated GaAs is found at

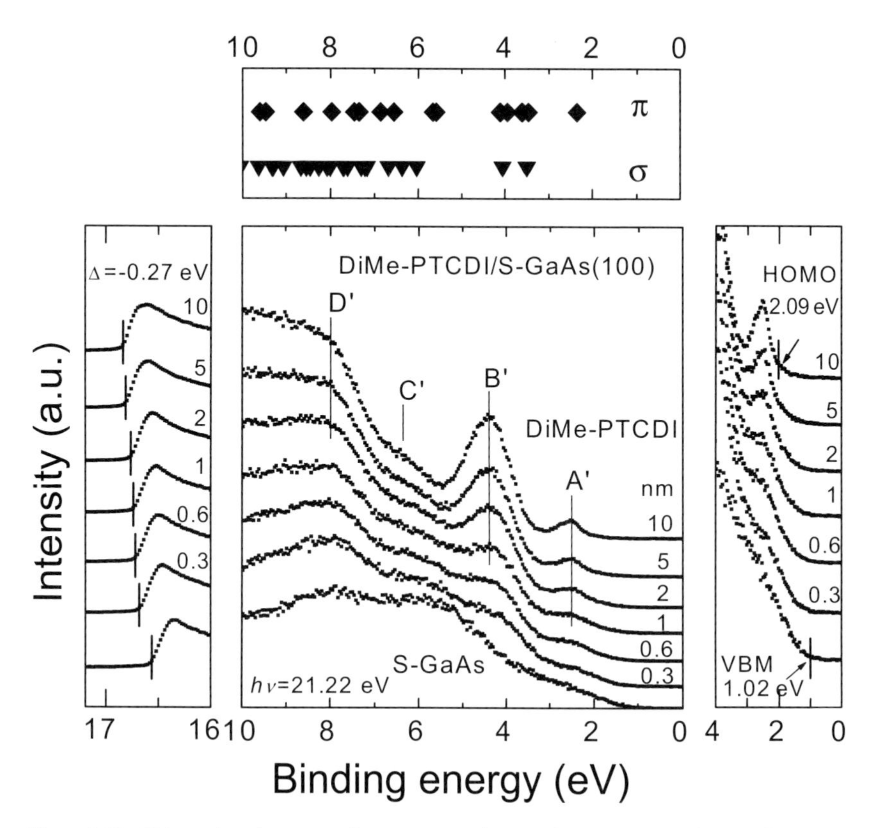

Figure 5.10 Valence band spectra of DiMe-PTCDI deposited on S-passivated GaAs for several thicknesses. Upper panel shows calculated energy positions of occupied π- and σ-orbitals. The spectra were measured with an He-discharge lamp ($h\nu = 21.2\,eV$).

(1.02 ± 0.10) eV with respect to the Fermi level E_F, while the ionization is determined to be (5.66 ± 0.10) eV. The evolution of the spectra with increasing DiMe-PTCDI deposition reveals that there is no detectable change in band bending in the GaAs substrate. The DiMe-PTCDI HOMO does not change its energy position as a function of the film thickness. As in the case of PTCDA films grown on Se-passivated GaAs(100) surfaces, there is no indication for "band-bending"-like behavior in DiMe-PTCDI. After 10 nm of DiMe-PTCDI, the HOMO position is (2.09 ± 0.10) eV relative to E_F. Therefore, the HOMO is located at (1.07 ± 0.10) eV below VBM. Moreover, the formation of an interface dipole can be derived from the shift of the secondary electron cut-off at high-binding energy (see the left panel of Figure 5.10). Here, there is a slow evolution of the low energy cutoff with coverage, while in Figure 5.7 there is a rapid shift for the lowest coverage. This different behavior is due to different growth modes of the organic materials on S-passivated GaAs(100). While the growth of PTCDA follows the Stranski–Krastanov mode, the growth of DiMe-PTCDI seems to start immediately with the formation of islands. For a nominal coverage of 0.3 nm, PTCDA has formed a monolayer and covers the whole substrate surface, while DiMe-PTCDI covers only a part of the substrate surface.

With the interface dipole of $\Delta = (-0.27 \pm 0.10)$ eV and the HOMO position, the energy level alignment diagram at the DiMe-PTCDI/S–GaAs(100) interface is determined and presented in Figure 5.11 in comparison with the energy level alignment of the PTCDA/S–GaAs(100) interface. For another sample, the ionization energy of the substrate was determined to be (6.28 ± 0.10) eV, which is close to the value of Se-passivated GaAs(100) samples. On this substrate, the interface dipole amounts to $\Delta = (-0.68 \pm 0.10)$ eV, which is larger than for the PTCDA/Se–GaAs(100) interface. The data for these two samples are also presented in Figure 5.9. A least square fit to the data gives an electron affinity for DiMe-PTCDI of (3.86 ± 0.10) eV. With the ionization energy of (6.46 ± 0.10) eV, a transport gap of (2.6 ± 0.10) eV is determined. Therefore, the LUMO of DiMe-PTCDI should be located 0.11 eV above E_{CBM}.

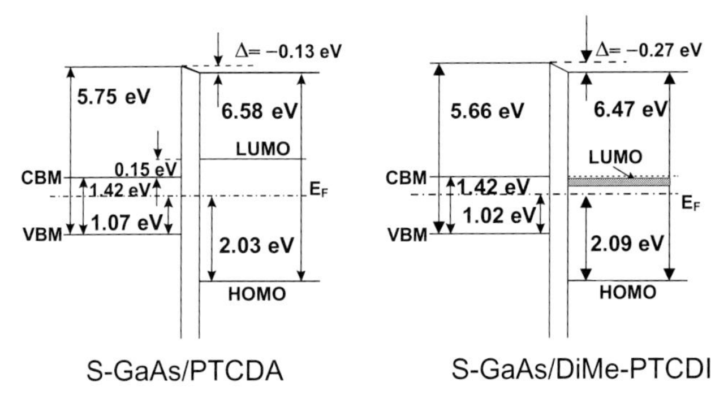

Figure 5.11 Energy level alignment at the interface between S-passivated GaAs(100) and PTCDA (left) and DiMe-PTCDI (right). Band bending in the substrate is not shown.

The electronic properties of inorganic semiconductor interfaces can be described within the interface-induced gap states model. These states change from donor- to acceptor-type character at the CNL. Once the CNL is determined for a semiconductor, it can be used to predict the energy level alignment at the semiconductor's interfaces. It is interesting to explore this concept toward its application to organic semiconductor interfaces.

Using the photoemission data presented here, a CNL of PTCDA and DiMe-PTCDI can be determined as follows. To determine a CNL for PTCDA, the PTCDA/GaAs(100)-c(4 × 4) interface will be considered, the energy level alignment at this interface being redrawn in Figure 5.12a. Compared to the other

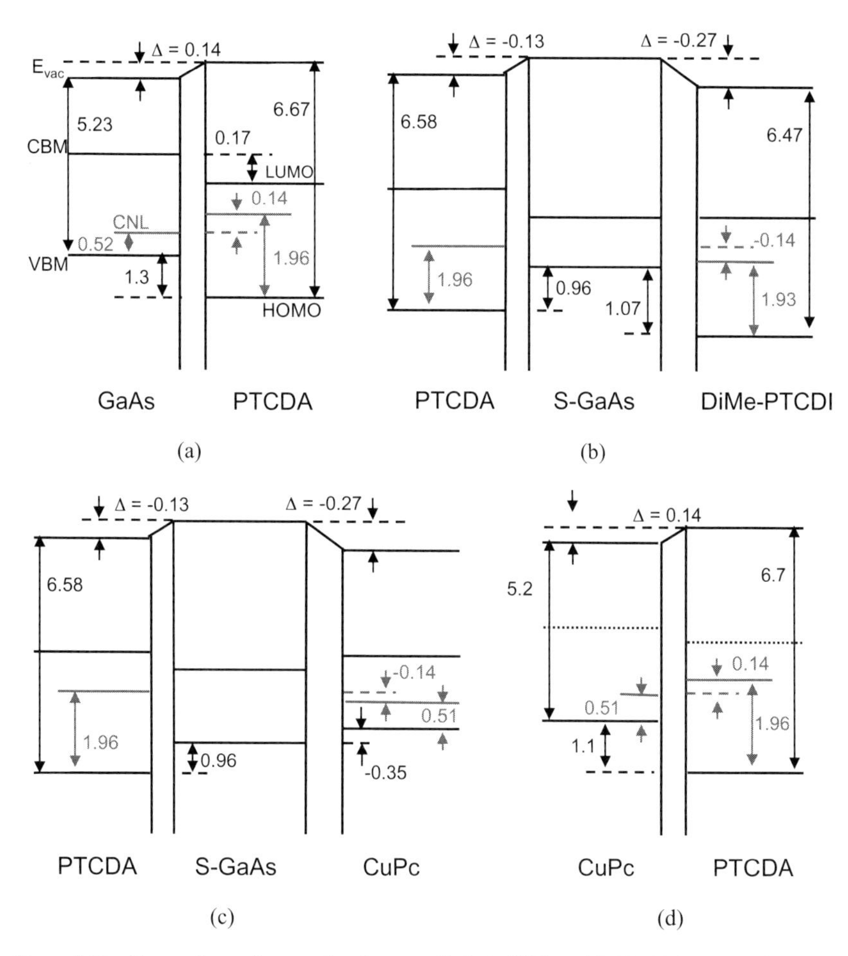

Figure 5.12 Comparison of energy level alignment at organic/inorganic semiconductor and organic/organic interfaces: (a) PTCDA/GaAs(100)-c(4 × 4); (b) PTCDA/S–GaAs(100) and DiMe-PTCDI/S–GaAs(100): (c) PTCDA/S–GaAs(100) and CuPc/S–GaAs(100), (d) PTCDA/CuPc. The positions of the respective CNLs are indicated in red.

GaAs(100) surfaces used in this study, the c(4 × 4) surface is the one closest to an ideally terminated GaAs(100) surface. Therefore, the position of the CNL is assumed to be 0.52 eV above the VBM, which is the value calculated for GaAs by Mönch using an empirical tight-binding approach. The CNLs on both sides of an ideal semiconductor–semiconductor interface should only differ by the interface dipole, which amounts to 0.14 eV for the PTCDA/GaAs(100)-c(4 × 4) interface. Accordingly, the CNL of PTCDA is 0.14 eV above the CNL of GaAs and its energy position is determined to be (1.96 ± 0.1) eV above the HOMO using the VBM–HOMO offset of 1.3 eV. To compare this value with the theoretical value of 2.45 eV determined by Vázquez *et al.* [227], it has to be taken into account that the experimental value is given with respect to the leading edge of the HOMO. From the photoemission spectra presented in Figure 5.7, the energy difference between the maximum of the HOMO feature and its leading edge is found to be 0.55 eV. This results in an energy position of the CNL of (2.41 ± 0.1) eV above the maximum of the HOMO feature, which is in excellent agreement with the theoretical results.

The CNL of DiMe-PTCDI can now be estimated by comparing the energy level alignment at PTCDA/S–GaAs(100) and DiMe-PTCDI/S–GaAs(100) interfaces as shown in Figure 5.12b. The VBM is taken as a reference level, and the slightly different Fermi level positions and ionization energies on the S-passivated GaAs(100) substrates (see Figure 5.11) are neglected.

Since the substrate is of the same type, the CNL of PTCDA and DiMe-PTCDI should differ by the difference in interface dipoles. Here, the interface dipole at the PTCDA/S–GaAs(100) interface is smaller, which positions the CNL of DiMe-PTCDI 0.14 eV below the CNL of PTCDA. Using the VBM–HOMO offsets at both interfaces, the CNL of DiMe-PTCDI is (1.93 ± 0.15) eV above the HOMO.

The S-passivated GaAs(100) was also used for the deposition of CuPc [235]. CuPc adsorption generates states above the GaAs VBM, changing the Fermi level position and causing a broadening of the core level emission peaks. While the passivating nature of the surface is unaffected by PTCDA adsorption on n-type GaAs(100), it is enhanced for CuPc adsorption. The PTCDA/S–GaAs(100) interface has a straddled band offset profile, while the CuPc/S–GaAs(100) interface has a staggered profile. The comparison of both interfaces, PTCDA/S–GaAs(100) and CuPc/S–GaAs(100), is shown in Figure 5.12c. Taking into account the difference in interface dipoles and the VBM–HOMO offsets, the CNL in CuPc is found to be (0.51 ± 0.1) eV above the HOMO.

The validity of the concept of CNLs at organic interfaces is checked by analyzing the energy level alignment at CuPc/PTCDA interfaces determined by Hill and Kahn [236]. Their results from PES investigations are shown in Figure 5.12d and give an interface dipole of 0.14 eV with the vacuum level of PTCDA being above the vacuum level of CuPc. Using the positions for the CNL of both organic materials, they are found to differ by exactly the interface dipole.

A final comment should be made concerning the CNL of the S-passivated GaAs(100) surface. In principle, it can be determined with the help of the CNL of PTCDA from the energy level alignment at the PTCDA/S–GaAs(100) interface. One obtains a position of 1.13 eV above the VBM. Compared to the CNL of GaAs shown in Figure 1.16, the CNL of the S-passivated surface should be at lower barrier heights on n-type GaAs. It should also be shifted toward the high-electronegativity metals like Pt or Pd, since sulfur is more electronegative than As resulting in a higher electronegativity of the S-passivated GaAs(100) surface. This trend is in principle supported by the barrier heights of metal contacts on S-passivated GaAs(100) surface, which are lower than on the nonpassivated GaAs(100) surfaces. The experimentally observed shifts in the barrier heights are smaller than the estimated shift in CNL. This can be due to secondary mechanisms like interface dipoles or defects.

The injection of carriers into organic semiconductors is a key process for the operation of organic-based devices, and is determined by the electronic and chemical properties of metal/organic interfaces. While low-workfunction materials are used for electron injection, metals with a large workfunction are used for hole injection. In many cases, the carrier injecting electrodes are prepared by evaporating metals on organic layer. These metal-on-organic interfaces are different from organic-on-metal interfaces in that metallization chemistry and interdiffusion can occur [224, 237, 238].

Here, results from PES, X-ray diffraction (XRD), and Raman spectroscopy investigations on the interfaces between metals and perylene derivatives will be presented. With Ag and In, two metals showing different interface reactivity will be used. PTCDA and DiMe-PTCDI films serve as substrates with different morphology and molecular orientation.

5.1.3
PTCDA/DiMe–PTCDI Metal Interfaces

Figure 5.13 shows valence band spectra of clean PTCDA and DiMe-PTCDI, and after successive Ag depositions. At Ag coverages below ~1 nm, all features from the PTCDA and DiMe-PTCDI films can still be seen. The lineshapes and energy positions of these features are not changed, indicating that the chemical interaction between Ag and the organic materials is very low and there is no Ag-induced band bending in PTCDA and DiMe-PTCDI. This is in agreement with the results by Hirose *et al.*, who found that the lineshape of the C1s core level emission of PTCDA is not changed upon deposition of 4 nm of Ag [224]. In addition, two small features appear in the HOMO–LUMO gap of the organic films (R1′,R2′ and R1″, R2″). The energy positions of those features are at 0.5 and 1.7 eV below the Fermi edge. The Fermi edge becomes visible above 2 nm. This thickness is four times larger than the thickness for which a Fermi edge is observed for Ag films grown on chalcogen-passivated GaAs(100) surfaces (0.5 nm). This can be attributed to the formation of clusters, which show metallic behavior at higher nominal coverages compared to epitaxial films.

While the features of the organic films become attenuated upon further Ag deposition, the new features in the HOMO–LUMO gap become stronger and the emission from the Ag4d band appears between 4~7.5 eV. The R1′, R2′, R1″, and R2″ features are often called gap states and have been observed for interfaces like PTCDA on Ag [239], or In, Sn, Al, and Ti on PTCDA [224], or Ag on 3,4;9,10-PTCBI [222]. Since the chemical interaction between Ag and PTCDA and DiMe-PTCDI,

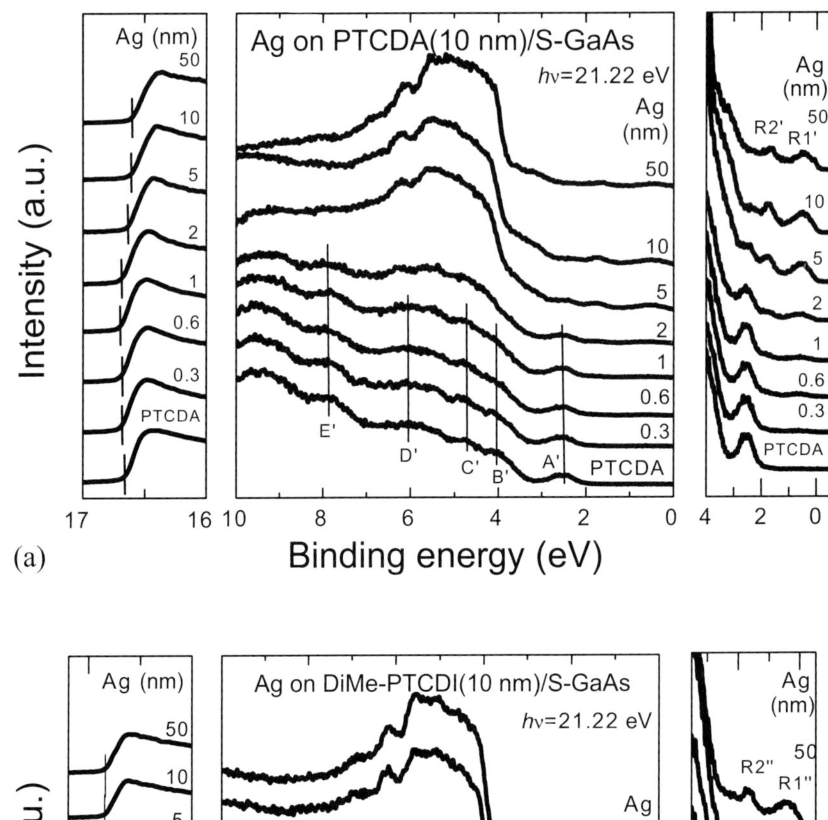

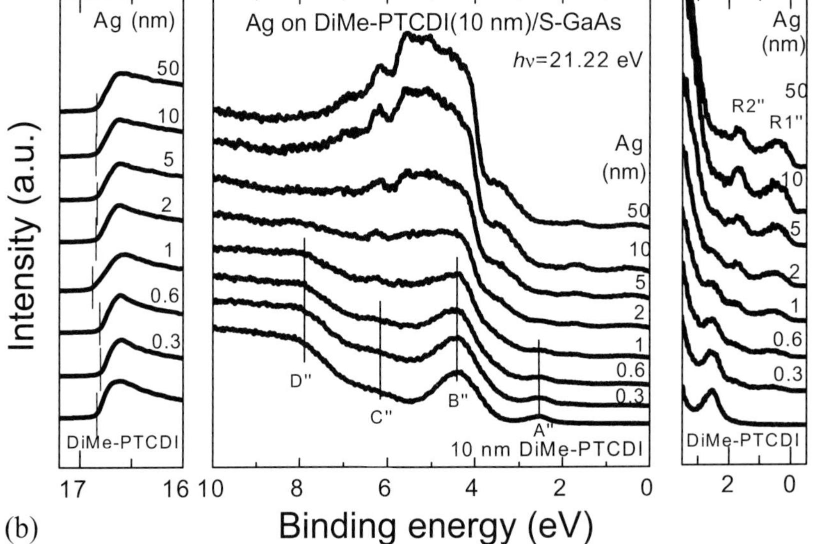

Figure 5.13 VB spectra as a function of Ag thickness on (a) 10 nm PTCDA and (b) 10 nm DiMe-PTCDI films ($h\nu$ = 21.22 eV). Solid lines are guides for the eye.

respectively, has been found to be very low, these gap states are attributed to a charge transfer between the metal atoms and the molecules resulting in polaron and bipolaron levels.

The bipolaron is occupied with two electrons and energetically positioned below the Fermi level. The polaron is occupied with one electron and should be positioned at the Fermi level. Since organic molecules represent small systems with localized charge electrons, the Coulomb repulsion between the first and second electrons placed in this orbital results in the opening of a Coulomb gap, such that the singly occupied polaron state will occur below the Fermi level [222]. In this sense, the states R1′, R1″, and R2′, R2″ are attributed to polaron- and bipolaron-like states, respectively. The R1′ and R2′ (R1″ and R2″) states can still be observed for a fairly high coverage with a maximum in intensity for a nominal coverage of 10 nm. As will be shown, the Ag films grown on PTCDA and DiMe-PTCDI are polycrystalline. These findings and the fact that the density of states of Ag is very low in the binding energy range where the R1′ and R2′ (R1″ and R2″) states appear lead to the low attenuation of the features as a function of the nominal coverage.

The evolution of the workfunction as a function of the thickness of Ag is shown in Figure 5.14a. The values for single crystal Ag(111), Ag(110), and Ag(100) and polycrystalline Ag films are included [240, 241]. Before Ag deposition, the ϕ values of PTCDA and DiMe-PTCDI are (4.56 ± 0.05) and (4.38 ± 0.05) eV, respectively. Upon deposition of Ag on PTCDA, a slight decrease in ϕ is observed up to 1 nm and then ϕ begins to increase gradually to reach a final value of (4.61 ± 0.05) eV. On the other hand, ϕ does not change overall for Ag deposition upon DiMe-PTCDI. The decrease in workfunction for Ag deposition on PTCDA can be explained by taking into account the formation of polaron and bipolaron states. The charge transfer between the metal atoms and the molecules results in the formation of molecular ions. Since Ag adsorbs on the organic films, these ions are positioned at the surface resulting in a local electric field, which changes the workfunction of the sample. From the subtle decrease in workfunction observed for PTCDA, an electron transfer from the metal to the molecule can be concluded.

Additional XRD experiments on the samples reveal two peaks corresponding to the (111) and (200) directions with the relative intensity ratios of the two peaks being different depending on the organic substrates (see Figure 5.14b). The (111) peak is much stronger for Ag/PTCDA than Ag/DiMe-PTCDI while the (200) peak intensities are similar in both cases. Therefore, the crystalline structure of the Ag films is strongly affected by the morphology of the underlying organic film.

The difference in ϕ for thick Ag films on PTCDA and DiMe-PTCDI can be understood by comparison with the ϕ values of Ag with different crystalline structures. The ϕ value of the Ag film on PTCDA is closer to the value corresponding to Ag(111) single crystal. This agrees very well with the observation of a stronger (111) peak in the XRD experiment.

Figure 5.15 shows valence band spectra during deposition of In on 20 nm films of PTCDA and DiMe-PTCDI. Unlike Ag, the deposition of the smallest amount

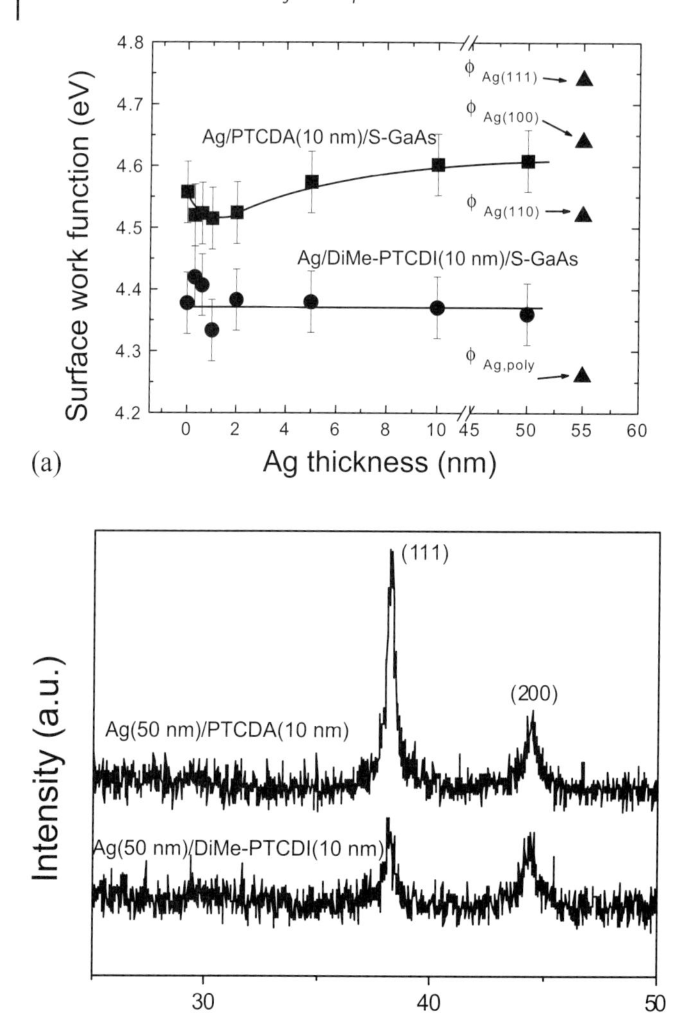

Figure 5.14 (a) The evolution of ϕ as a function of Ag thickness deposited on 10-nm films of PTCDA and DiMe-PTCDI. For comparison, values of ϕ for the polycrystalline Ag film and single crystalline Ag(111), Ag(1101), and Ag(100) are included. Solid lines are guides for the eye. (b) XRD spectra of 50 nm Ag films on the 10 nm PTCDA and DiMe-PTCDI films.

of In (0.1 nm for PTCDA and 0.05 nm for DiMe-PTCDI) changes the spectra significantly, namely, the appearance of two new features (R3′, R4′ and R3″, R4″) in the HOMO–LUMO gap of the organic films and strong energy shifts are observed in the low kinetic energy onset as well as in the bands A″ and B″ of DiMe-PTCDI. Due to the deposition of In, the features in the photoemission spectra are broadened. As a result, the features in the PTCDA almost disappear and energy shifts cannot be determined correctly. The changes in the spectra are attributed to a

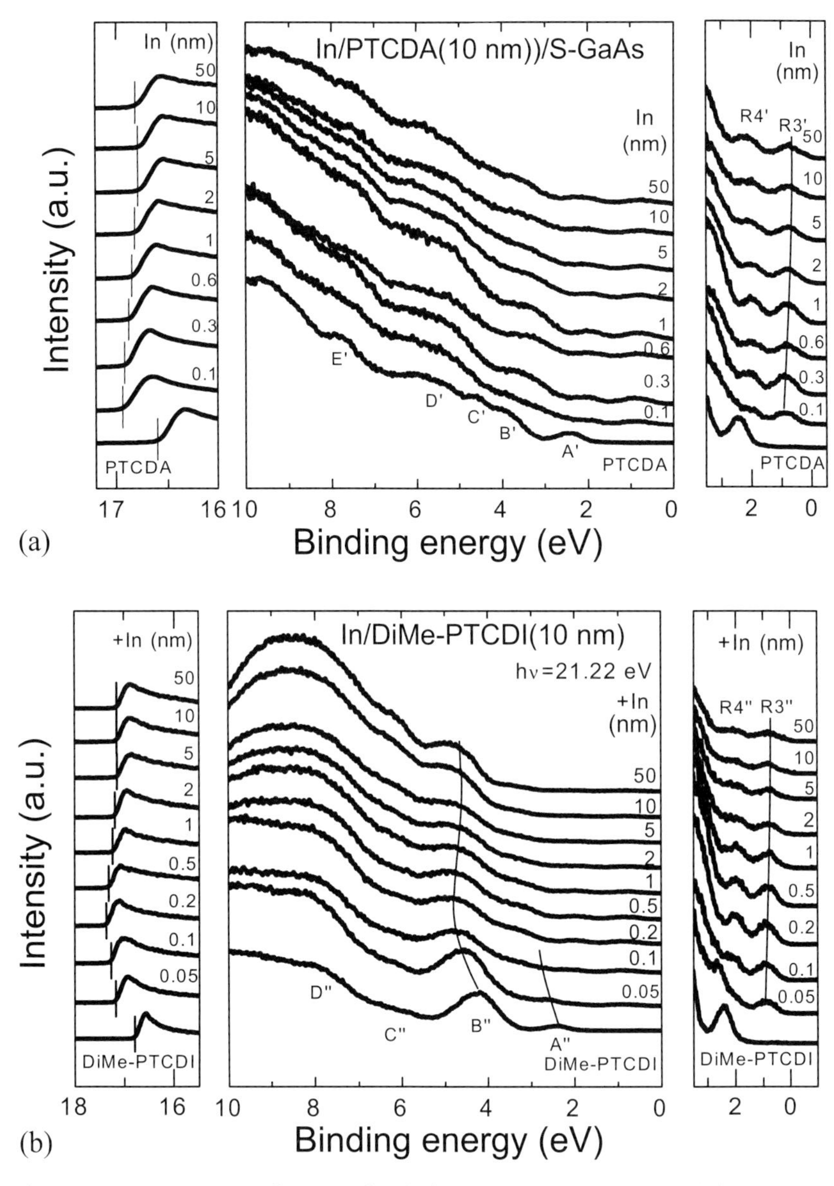

Figure 5.15 VB spectra as a function of In thickness on (a) 10 nm PTCDA and (b) 10 nm DiMe-PTCDI films ($h\nu = 21.22$ eV). Solid lines are guides for the eye.

pronounced interaction between In and PTCDA, which may be accompanied by a strong diffusion of In atoms into the PTCDA crystallites. However, for In/DiMe-PTCDI, the shifts of A″ and B″ toward higher binding energy are unambiguous and the band B″ is visible up to 50 nm In thickness. The new states in the gap are different from those which appear upon Ag deposition (R1′,R2′ and R1″,R2″ in Figure 5.13) in that they are pronounced even at low In thickness. These new states

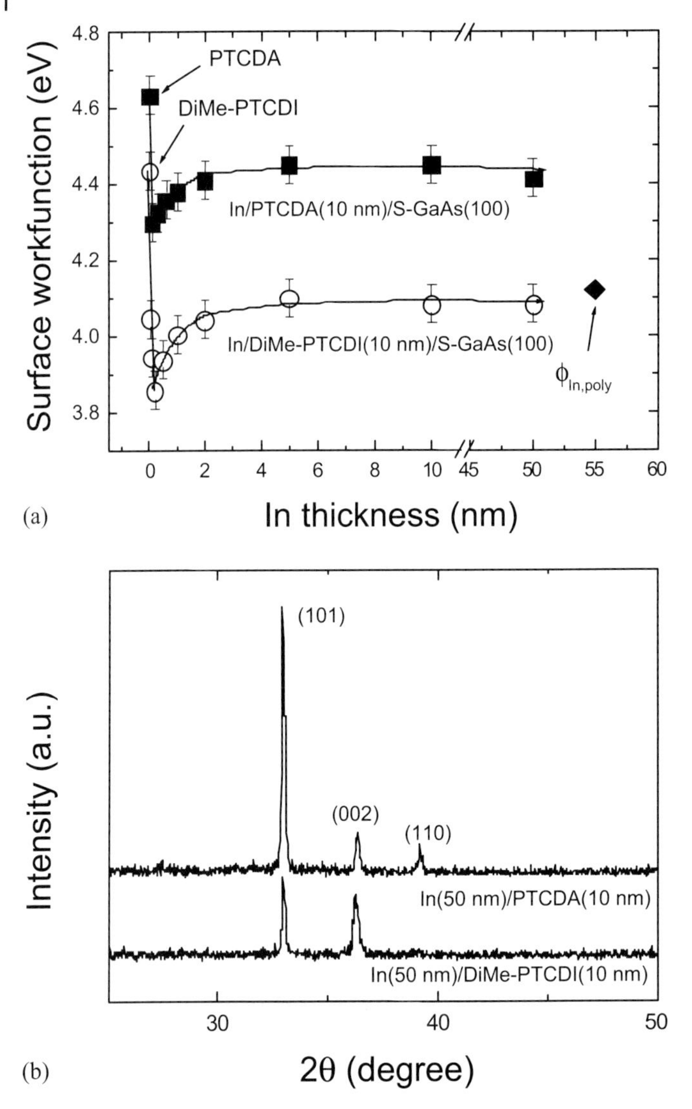

Figure 5.16 (a) The evolution of ϕ as a function of In thickness deposited on 10 nm films of PTCDA and DiMe-PTCDI. For comparison, the value of ϕ for the polycrystalline Ag film [Pei71] is included. Solid lines are guides for the eye. (b) XRD spectra of 50 nm In films on the 10 nm PTCDA and DiMe-PTCDI films.

are again attributed to the formation of polaron- and bipolaron-like states. Compared to Ag, In seems to diffuse into the organic films. Therefore, molecular ions are formed within the organic film.

The evolution of ϕ is shown in Figure 5.16a as a function of In thickness on 10 nm films of PTCDA and DiMe-PTCDI. A very steep decrease in ϕ at low In thickness was observed for both molecules, followed by a gradual increase up to final values. As in the case of Ag, the different final values of ϕ can be attributed to different crystalline structure and crystallinity of the In layers.

The differences between ϕ of the bare organic films and the minimum ϕ at low In thickness are (0.40 ± 0.10) and (0.55 ± 0.10) eV for In/PTCDA and In/DiMe-PTCDI, respectively. The charge transfer between In atoms and organic molecules is apparent from these changes in workfunction for low metal coverage. In both cases, a charge transfer from the metal to the molecule takes place. Due to the formations of molecular ions in the organic films, the polarization energy is changed, which changes the energy position of the molecular orbitals. This can be observed for the highest intensity feature B″ in the DiMe-PTCDI spectra, which shifts to higher binding energies. On the other hand, the diffusion of In may induce a gradient in the concentration of molecular ions in the organic films. Accordingly, the changes in polarization energy and binding energies should vary, resulting in a broadening of the features observed in the photoemission spectra.

The XRD measurements for the 50 nm In film on PTCDA and DiMe-PTCDI (Figure 5.16b) also show different morphology. For the In film on PTCDA, three peaks corresponding to the (101), (002), (110) crystallographic directions are revealed, while only (101) and (002) peaks are visible for In on DiMe-PTCDI. The intensity of the (101) peak is found to be much stronger for In/PTCDA. Considering also the stronger (111) diffraction peak for Ag/PTCDA than that for Ag/DiMe-PTCDI, it can be concluded that metal films on PTCDA have higher crystallinity than on DiMe-PTCDI. The interaction seems to be strongest for the In/PTCDA interface.

5.1.4 Bandstructure of PTCDA and DiMe-PTCDA

Inorganic semiconductors the energy dispersion of the valence band has an energy width of a few electron Volt due to the strong covalent/ionic bond formation between next-neighbor atoms. In organic semiconductors, next neighbour interaction is mostly of the much weaker van-der-Waals type, and energy band dispersions are expected to be about 100 meV wide. Due to the anisotropic nature of organic materials, these band dispersions may only be observable in particular directions. In the case of PTCDA and DiMe-PTCDI, strongest interaction is expected in the direction of maximum π electron overlap, that is, perpendicular to the molecular plane. This has some implications for the geometry, which has to be chosen to investigate band dispersion. Especially in the case of PTCDA, which mostly grows with its molecular plane parallel to the substrate surface, the strongest molecular interaction along the surface normal. Therefore, angular resolved photoemission measurements along directions parallel to the substrate surface may not be helpful. Instead, the direction perpendicular to the substrate surface is of particular interest. In this measurement geometry, which is called normal emission, the wave vector has only a component $k_\perp$ perpendicular to the surface and $k_\perp$ is scanned by changing the excitation energy.

Measurements on band dispersion in organic materials have been performed on π-conjugated polymers [242, 243, 244] and on small molecules like C_{60} [245]. The width of the valence band, or, in other words, the intermolecular energy dispersion in the HOMO has been found to be in the order of a few tenth of an electron Volt. Yamane and coworkers have investigated the intermolecular energy dispersion in PTCDA thin films [246]. The PTCDA had been deposited onto MoS_2 where it grows in nicely ordered films with different domains and the molecular plane being parallel to the substrate surface. An energy dispersion in the HOMO of 0.2 eV is indeed found along the surface normal. Gavrila and coworkers [247] applied the same experimental procedure to DiMe-PTCDI thin films. Their data as well as the data evaluation procedure employed by both groups will be presented in more detail now.

An energy dispersion along $k_\perp$ is accessible by changing the photon energy stepwise and looking for periodic shifts in the binding energy of the valence states. Such a measurement can only be performed at synchrotron radiation facilities, where high resolution beamlines provide access to tuneable VUV and soft X-ray light. Figure 5.17. shows a data set taken from a 15 nm DiMe-PTCDI film grown on sulphur passivated GaAs(100). The spectra have been taken for different photon energies and are plotted as a function of the binding energy relative to the vacuum level. The well resolved HOMO and HOMO-1 features at 7 eV and 9 eV, respectively, show a periodic shift in binding energy as a function of the photon energy. An $E(k_\perp)$-plot is obtained from this data set by the following procedure.

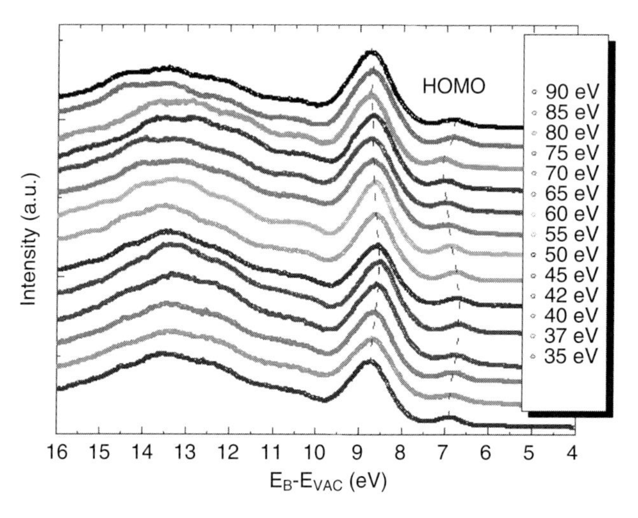

Figure 5.17 Photoemission spectra of a DiMe-PTCDI film using photon energies in the range of 35eV–90 eV.

The energy and momentum conservation in (5.1) and (5.2) may be rewritten as

$$E_f = h\nu + E_i,\ k_f^{\perp} = k_i^{\perp} + G^{\perp} \tag{5.4}$$

with initial and final electron energies E_i and E_f relative to the vacuum level. The momentum of the photon is neglected in the momentum conservation law and $k_i^{\perp}$ and $k_f^{\perp}$ differ by a reciprocal lattice vector $G^{\perp}$. In a first order approximation the final state above the vacuum level is assumed to be free electron like with a parabolic energy dispersion:

$$E_f = \hbar^2 k^2 / 2m^* + V_0, \tag{5.5}$$

where m^* is the effective mass of the electron and V_0 is the constant inner potential in the solid for the final state. Since electron energies are given with respect to the vacuum level, the final state energy corresponds to the kinetic energy of the electrons and (5.4) and (5.5) can be rewritten as

$$E_i = E_{kin} - h\nu \tag{5.6}$$

and

$$k_i^{\perp} = k_f^{\perp} = [2m^* \times (E_f - V_0)]^{1/2} / \hbar = [2m^* \times (E_{kin} - V_0)]^{1/2} / \hbar \tag{5.7}$$

where the effective mass m^* of the excited electron is approximated by the mass of a free electron m_0. Photoemission spectra are recorded as a function of the kinetic energy of electrons and a change in photon energy would shift the spectra by the same amount. Those shifts are compensated by plotting the spectra as a function of the binding energy relative to the vacuum level and energy dispersions related shifts will be revealed. With E_B being the binding energy relative to the vacuum level (5.7) changes to

$$k^{\perp} = [2m^* \times (h\nu - E_B - V_0)]^{1/2} / \hbar. \tag{5.8}$$

Under the assumption that only next neighbor interaction has to be taken into account the energy dispersion $E_B(k^{\perp})$ of the HOMO can be approximated in a one-dimensional (1D) tight binding approach [248]:

$$E_B(k^{\perp}) = E_B^0 + 2t\cos(a^{\perp} \times k^{\perp}) \tag{5.9}$$

where t the transfer integral describing the strength of next neighbor interaction. The periodicity along the surface normal $a^{\perp}$ may be determined from structure measurements. From Raman and IR investigations DiMe-PTCDI molecules/molecular planes are found to be tilted by 56° with respect to the substrate surface. With a 3.21 Å spacing between two adjacent molecular planes the lattice spacing normal to the surface amounts to $a^{\perp}$ = 3.9 Å.

The energy dispersion for the HOMO in PTCDA and DiMe-PTCDI films are shown in Figure 5.18. The data sets are aligned along the momentum direction. Both data sets show a periodic shift in the binding energy of the HOMO, that is, an energy dispersion. However, amplitude and periodicity are different. A fit to the data using (5.8) and (5.9) gives a transfer integral $t = 0.04$ eV for both molecules,

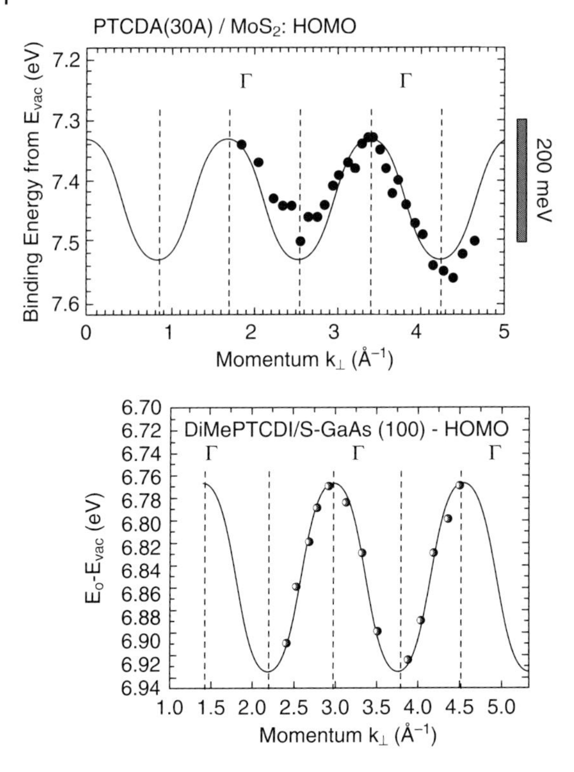

Figure 5.18 Experimental HOMO dispersion of PTCDA and DiMe-PTCDI thin films. Lines show results of curve fitting using a tight binding model.

while the inner potential V_0 amounts to−5.3 eV and and −5.1 eV for DiMe-PTCDI and PTCDA, respectively. In addition the effective mass of the HOMO hole can be determined from the second derivative of the HOMO energy dispersion. The value $m_h^* = 6.20\,m_0$ determined for DiMe-PTCDI films is larger than the value of $5.28\,m_0$ determined for PTCDA.

The different periodicity in the band dispersion of PTCDA and DiMe-PTCDI is a direct consequence of the different orientation of the molecular thin films with respect to the substrate surface. As shown in Figure 5.19 PTCDA molecules lie flat on the surface and the highly ordered thin film is oriented with the (102) direction being almost parallel to the surface normal. In other words, the measurements are performed along the direction of maximum π-electron overlap, which results in a bandwidth of about 200 meV. In the case of DiMe-PTCDI the (102) direction is significantly tilted with respect to the surface normal. The effective distance between two successive layers is larger, resulting in an smaller π-electron overlap and therefore a smaller bandwidth of about 150 meV. In addition, the curvature of the energy dispersion is reduced and resulting in a larger effective hole mass. Compared to the HOMO the energy dispersion in the HOMO-1 is less

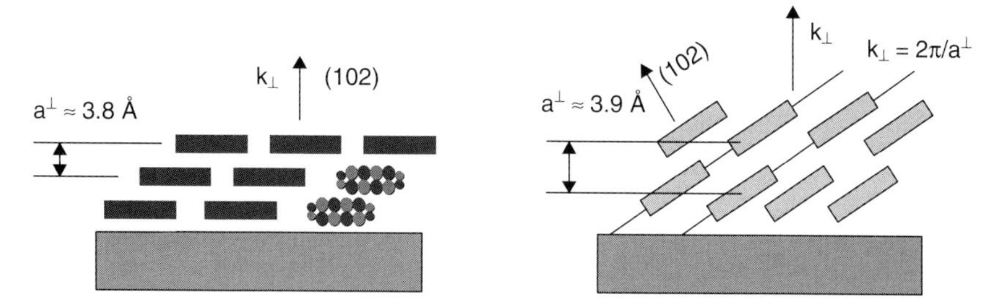

Figure 5.19 Orientation of the PTCDA and DiMe-PTCDI molecules and layers with respect to the substrate surface.

pronounced. This is explained by the higher binding energy and the predominant σ character of the molecular orbitals contributing to HOMO-1.

5.2 Inverse Photoemission

IPES may be seen as the time reversal of PES. In IPES, an electron is injected into the solid on an energy level E_i above the vacuum level. The electron relaxes into a lower energy state below the vacuum level and the excessive energy is released as a photon. The emitted photon is detected by a Geiger–Müller tube, where it ionizes a gas mixture. The electron source and the photon detector typically operate in the 10 eV range. The kinetic energy for the incident electron is varied, while the detector detects photons at a fixed energy. In the detector, the optical adsorption edge of the window and the ionization energy of the gas mixture set a band pass for the detectable photon energy. For an MgF window and ethanol, the energy window is at 10.9 eV. The overall resolution of such an experimental setup is about 0.4 eV. It should be mentioned that in Bremsstrahlung isochromat spectroscopy, photons of about 1400 eV are detected.

The IPES process has a much lower yield than comparable photoemission events. The count rate from photoemission events will, therefore, be significantly reduced, about three to five orders of magnitudes.

Figure 5.20 shows the densities of unoccupied states (DUOS) of PTCDA and DiMe-PTCDI measured by IPES. The deconvolution of the experimental data was achieved by fitting the experimental spectra with Gaussian functions and then correcting the widths of the fitted peaks for the experimental resolution. These data are compared to the calculated DUOS, which is obtained by Gaussian broadening of the calculated energies of unoccupied orbitals. For both molecules, the calculated data had to be shifted by about 0.8 eV toward the Fermi energy. Figure 5.20 demonstrates that the LUMO levels are not affected by the substitution of anhydride oxygen by methylamine groups. In both molecules, most of the states are predominantly delocalized over the perylene core.

Combining photoemission and IPES enables one to determine the band gap of the organic semiconductor, that is, the energy distance between HOMO and LUMO (see Figure 5.21). Compared to optical spectroscopies, no excitons are involved and the resulting band gaps should be closer to the ones of the neutral/nonexcited organic semiconductor. The energy position of the HOMO and LUMO is determined by extrapolating the spectral feature respective peak to higher and lower energies, respectively. The resulting energy difference is called the transport gap. The name is due to the fact that the charge transfer is carried out by the highest occupied and the lowest unoccupied states. An alternative approach is to

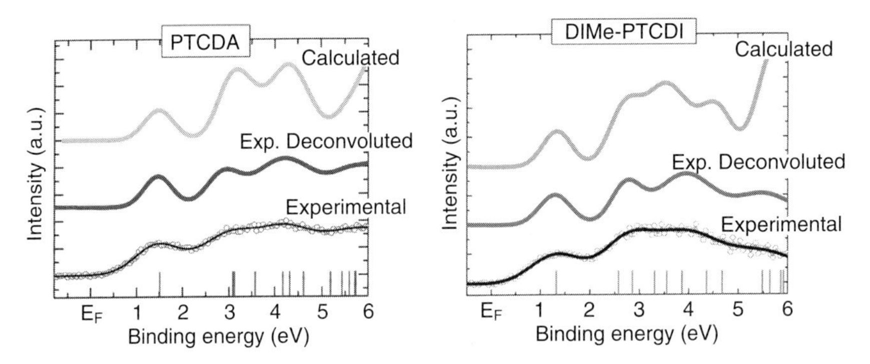

Figure 5.20 Experimental, deconvoluted, and calculated densities of unoccupied states of PTCDA and DiMe-PTCDI. [from Reference 232]

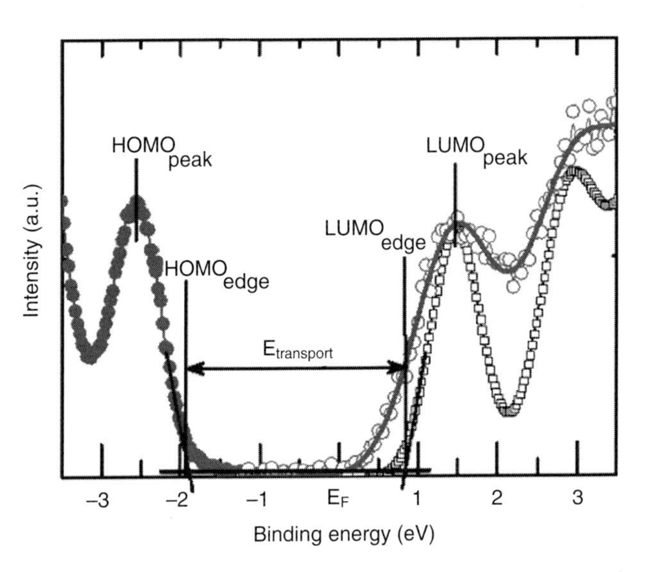

Figure 5.21 Combined density of occupied and unoccupied states of PTCDA determined by UPS and IPES, respectively. [from Reference 232]

use the respective peak positions of the HOMO and LUMO for the determination of the gap.

5.2.1 Band Gaps of Perylene Derivatives

Figure 5.22 shows the combined UPS and IPES data of PTCDA, PTCDI, and DiMe-PTCDI thin films grown on sulfur-passivated GaAs(001) by OMBD. The film thickness is about 15 nm. Contributions of the individual peaks to the overall

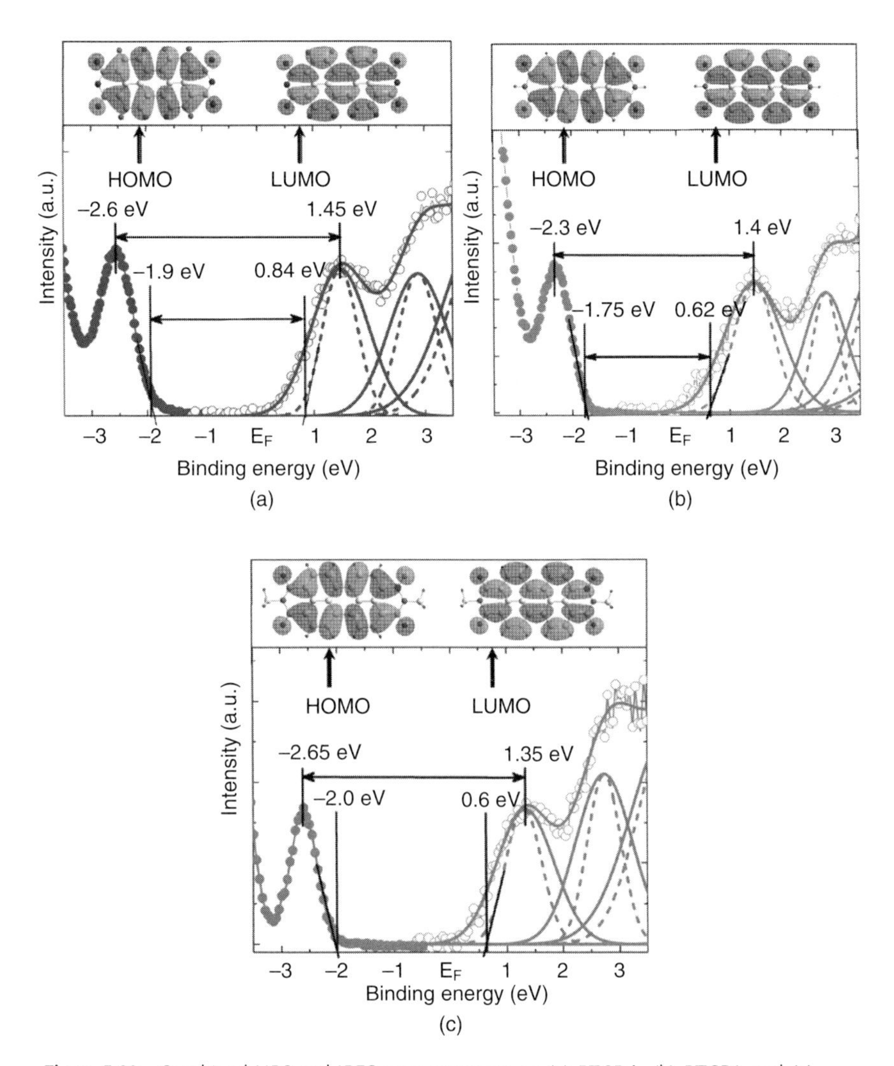

Figure 5.22 Combined UPS and IPES measurements on (a) PTCDA, (b) PTCDI, and (c) DiMe-PTCDI. [from Reference D. R. T. Zahn, G. N. Gavrila, M. Gorgoi, *Chem. Phys.* **325** (2006) 99.]

intensities of the experimentally measured IPES data are plotted with lines, while the deconvoluted contributions are plotted with dashed lines. The UPS and IPES spectra for all the perylene derivatives were aligned on the energy scale with respect to the Fermi level. The HOMO and LUMO energy positions (edges) are determined from the intercept of two linear extrapolations.

5.3 Total Current Spectroscopy

In total current spectroscopy (TCS) the density of unoccupied states above the vacuum level is determined by measuring the electron yield or sample current as a function of the energy of incident electrons. [249] In an elastic scattering model the TCS signal can be described by the energy dependence of the reflectivity coefficient for electrons. [250] If the primary electron energy overlaps with an energy gap in the unoccupied states, the electrons can not "couple" to the states in the sample and are reflected at the surface. If, on the other hand, the kinetic energy of the incident electrons overlaps with unoccupied states in the sample, electrons can enter the sample and reflectivity is low. At the boundary between a gap and a band, the strong decrease in electron reflectivity induces a maximum in a total current (TC) spectrum. Therefore, maxima in the TCS signal can be associated with the energy position of unoccupied band edges. [251, 252]

Inverse photoemission spectroscopy and TCS are sensitive to delocalised unoccupied states and may be considered to be complementary because of their respective energy ranges: IPES probes the density of unoccupied states between Fermi level and vacuum level, while TCS is sensitive to unoccupied states above the vacuum level. Near-edge X-ray absorption fine structure spectroscopy (NEXAFS), which will be explained in more detail in Section 5.4, is a more "local" probe when compared with IPES and TCS. It is sensitive to the partial density of states localized near the atom of which the core-level electron is excited.

One way of performing a TCS experiment is to use a slightly modified four-grid LEED system. Usually a four-grid LEED optic is used to do electron diffraction as well as Auger electron spectroscopy at kinetic electron energies well above a few electron Volts. In TCS very low electron energies starting practically from zero energy are used. In order to get a collimated electron beam at such low energies an additional retarding field is applied between the last electron gun electrode and the sample. The sample is grounded, and the primary electron energy is defined by the cathode bias potential.

PTCDA has been extensively studied using inverse photoemission (IPES) [253] and near-edge X-ray absorption fine structure spectroscopy (NEXAFS). [254] Here, TCS studies on PTCDA grown on H-passivated Si(111) are presented and discussed. [255] The H-passivation was achieved by wet chemical etching of the Si(111) samples in solutions containing HF. The data taken from the clean as well as the PTCDA covered substrate are shown in Figure 5.24. The TCS signal is plotted as a function of energy above the vacuum level.

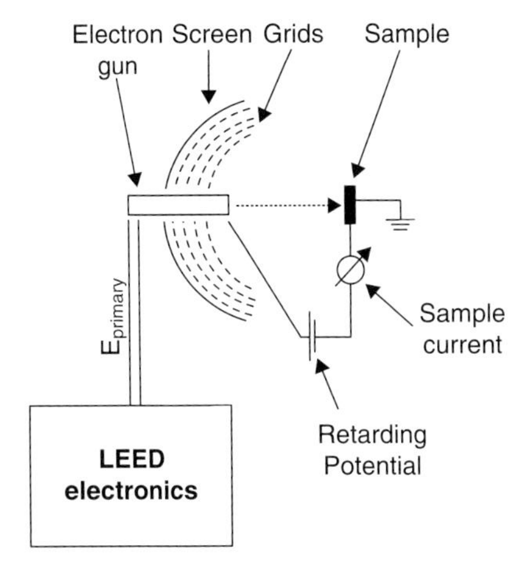

Figure 5.23 Experimental setup for total current spectroscopy.

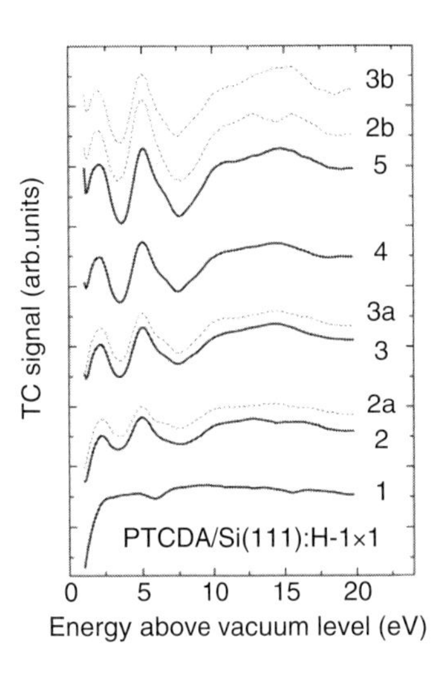

Figure 5.24 Evolution of the TC spectra during PTCDA film deposition. Solid lines represent experimental data of the substrate and for successive deposition times of PTCDA. 1: clean H-passivated Si(111) substrate; 2: 10 min deposition of PTCDA; 3: 20 min; 4: 30 min; 5: 60 min (~10 nm). 2a and 3a are the results of a linear combination of 1 and 5. 2b and 3b are the respective PTCDA contributions in 2a and 3a, respectively.

The spectrum for the passivated silicon surfaces lacks intensive fine structure and is very similar to that reported for a Si(111)-7 × 7 surface *in-situ* covered with hydrogen. [256] However, the data is different from data reported on the hydrogen-terminated silicon surface. [257] Since fine structure in TC spectra is known to be strongly dependent on the surface treatment [258] and on the 3-dimensional long-range ordering in the surface region, this difference can be attributed with the different etching procedures.

PTCDA induced features in TCS spectra are already observed for the lowest deposition time of 10 min, which corresponds to a film thickness of about 6 nm.

These features become more intense and distinct with increasing PTCDA coverage, but they do not change their energy position. Spectra do not change for deposition time larger than 60 min. This behavior can be explained by taking into account the growth mode of PTCDA on the H-passivated silicon, which is island growth as revealed by AFM. Therefore, the spectra shown for intermediate coverages are a superposition of the signal coming from the PTCDA islands and the uncovered H-passivated Si(111) substrate in between. These spectra can be simulated by a linear combination of spectra 1 and 5 for the clean substrate and the thick PTCDA film, respectively. The resulting spectra 2a and 3a agree well with the experimental counterpart. From this linear combinations the PTCDA spectra for the intermediate coverage can be determined. The results are shown as curve 2b and 3b in Figure 5.24. The TC spectra clearly show that already, for the lowest coverage, the TC spectra consist of the fine structure characteristic for a thick PTCDA film. This is due to the low molecule-substrate interaction and the immediate formation of islands. The intensive fine structure in the PTCDA spectra is typical for layered graphite-like structure, and has been observed with graphite, [259] VSe_2 and TiS_2, to name a few examples.

5.4 Near Edge X-ray Absorption Fine Structure Spectroscopy

In NEXAFS, absorption of the tuneable synchrotron radiation takes place by exciting electrons from core shells into unoccupied states above the Fermi level. Compared to PES, the excited electrons do not leave the solid. The excited electrons relax into the ground state and the excessive energy is released by the emission of either X-ray radiation or an Auger electron. Both can be used for signal recording, but one has to keep in mind that with Auger electrons, more surface sensitivity is achieved. In addition, one can also detect the photocurrent generated by the absorption of photons. In either way, the signal is detected as a function of the incoming photon energy. Since in the studies presented here electrons are excited from the K shells, photon energies of a few hundred electron volts are used. In the geometry shown in Figure 5.25, the arrow O_π indicates the direction of transition dipoles involved in an excitation from the K shell to the π^*-states. The absorption intensity can be expressed as a function of the angle between the electric field vector and the direction of the dipole transition.

Since the initial state is highly symmetric, angular dependent measurements reveal the symmetry of the final state. This is schematically shown in Figure 5.26. Here, a molecule is adsorbed on a surface and its unoccupied π^*-orbitals are parallel to the surface normal. For linear polarized under normal incidence, the electric field vector is perpendicular to the π^*-orbitals and no excitation into this orbitals will occur. Tilting the sample with respect to the incoming synchrotron light, the electric field will have a field component along the π^*-orbitals and the intensity of the respective π^*-resonance will increase. Considering π^* unoccupied states in

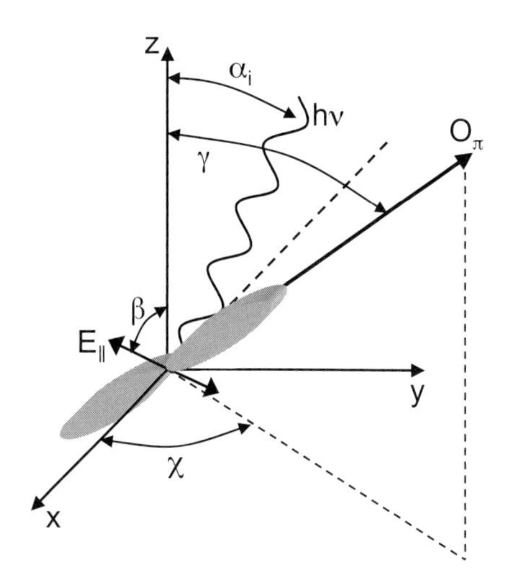

Figure 5.25 Geometry of a NEXAFS measurement.

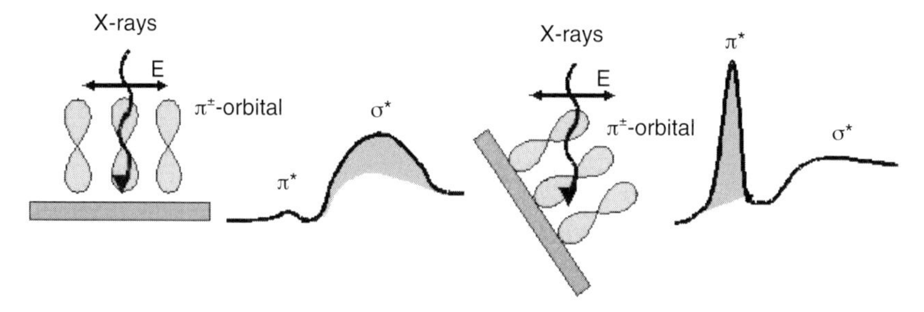

Figure 5.26 NEXAFS on a molecule with unoccupied π^*-orbitals using linear polarized light. The intensity of the π^*-resonance vanishes for the polarization vector **E** being perpendicular to the π^*-orbitals, intensity for **E** having a component along the π^*-orbitals.

perylene derivatives, maximum intensity is expected when the electric field vector of the synchrotron light is polarized perpendicular to the molecular plane, that is, parallel to the π^*-orbitals [258].

NEXAFS spectroscopy was performed at the PM1 and RGBL beamlines of the synchrotron light source BESSY II. The data were recorded in total yield mode and the light incidence angle α_i was varied between normal and near-grazing incidence, that is, between 0° and 70°. Spectra of a 100-nm thick Ag film were taken for normalization purposes.

Figure 5.27 presents selected C–K edge NEXAFS spectra of a 20-nm thick PTCDA film grown on S-passivated GaAs(100) for three different angles of incidence. The lower photon energy features near 280 eV correspond to π^*-states. Here, π^*-resonances have a negligible intensity for an incident angle of 0°. The

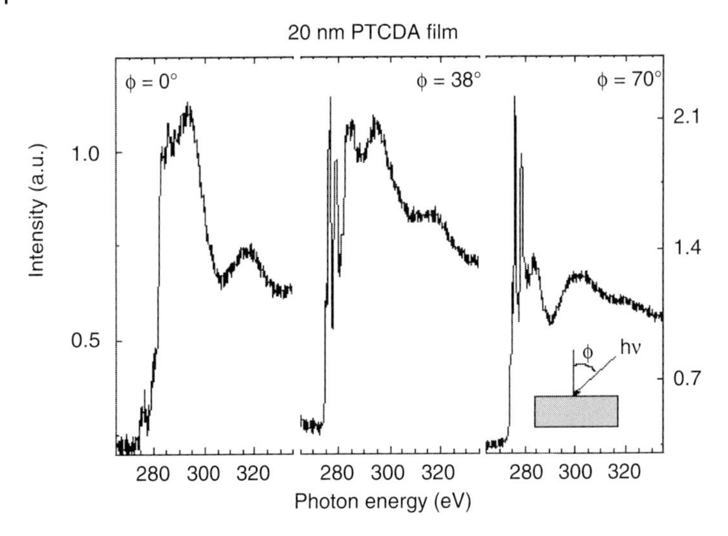

Figure 5.27 NEXAFS spectra taken from a 20-nm PTCDA film.

intensity increases with increasing incidence angle showing a maximum intensity for the largest angle of 70°. The increase in intensity of the π^*-resonances with increasing angle of incidence, which was also observed for the O–K edge, provides clear evidence for the parallel orientation of the molecular plane with respect to the substrate surface [259].

The same type of measurements was performed for DiMe-PTCDI grown on sulfur-passivated GaAs(001). Figure 5.28 shows a set of NEXAFS spectra taken under different angles of the incident linear polarized synchrotron light. It is evident from the nonvanishing intensity of the π^*-resonance at about 510 eV that the molecules within the thin film are not oriented parallel to the substrate surface. This becomes even more evident by plotting the intensities of the three most features in this energy range as a function of the incidence angle of the linear polarized synchrotron light. The π^*-resonance at 510 eV shows a maximum at an incidence angle of −20 ° with respect to the surface normal. This indicates that all molecules within this film possess a tilt into the same direction.

The tilt angle of the molecular plane θ and the projection χ of the long axis of the molecular plane onto the substrate can be obtained by performing angular dependent measurements for two different azimuthal orientations of the substrate. These measurements were performed on the CK-edge. The intensity ratio for the bands related to π^*-resonances is presented in Figure 5.29 (symbols) as a function of the incidence angle for two azimuthal directions of the sample: when the plane of incidence contains the [011] (Figure 5.29a) and [0–11] (Figure 5.29b) directions of the S-passivated GaAs(100) substrate. The simulation is performed using a model that assumes the same orientation for all molecules. With the electric field vector of the linear polarized synchrotron light being in the plane

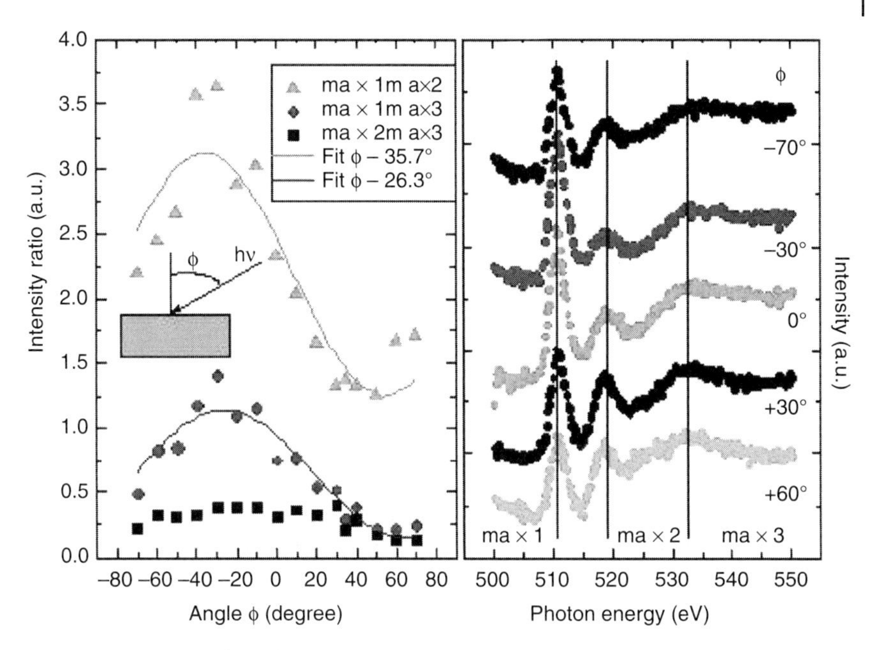

Figure 5.28 Angular dependent NEXAFS measurements on DiMe-PTCDI grown on sulfur-passivated GaAs(001) taken at the O K-edge. Spectra taken for different incidence angles of the linear polarized synchrotron light are shown on the right hand side, while the intensities of three distinct spectral feature as a function of the incidence angle are shown on the left hand side.

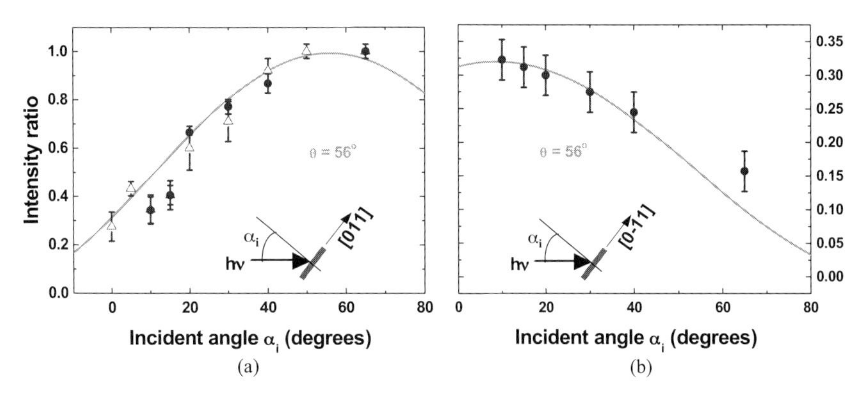

Figure 5.29 Experimental (symbols) and calculated (lines) intensity ratios for π^*-resonances as a function of the incident angle. Different symbols indicate data from different samples. Intensities are normalized to the maximum intensity of the π^*-resonance recorded for incidence angle of 50° in the [011] direction.

of incidence, the relationship between the NEXAFS intensity S and the angles describing the molecular orientation is obtained following the algorithm:

$$S = \cos^2\beta\cos^2\gamma + \sin^2\beta\sin^2\gamma\cos^2\chi + 2\sin\gamma\cos\gamma\sin\beta\cos\beta\cos\chi \qquad (5.10)$$

Here, β is the angle between the electric field vector of the polarized synchrotron light and the surface normal, and is given by $90° - \alpha_i$. The angles χ and γ are the azimuth and polar angle of the π-orbitals, respectively. The best fits to the experimental data are also presented in Figure 5.16. An average tilt angle of $\theta = 90° - \gamma = 56° \pm 5°$ of the molecular plane is obtained while for the in-plane projection an angle of $\chi = -7° \pm 3°$ with respect to the [011] axis of the GaAs substrate is evaluated. The orientation of DiMe-PTCDI molecules on S-passivated GaAs(100) will now be further evaluated by considering Raman spectroscopy results.

6
Charge Transport

Another physical property that is affected by the weak next-neighbor interaction and anisotropic crystalline structure is the carrier mobility. In PTCDA films, for instance, the conductivity is extremely anisotropic, with the in-plane conductivity being lower by at least six orders of magnitude compared to that perpendicular to the (102) plane.

The number of charge carriers n and their mobility μ determine the electrical conductivity σ of a material:

$$\sigma = ne_0\mu \tag{6.1}$$

where e_0 is the elementary charge. For electrons and holes, the charge will be negative and positive, respectively. The upper limits in microscopic mobilities of organic molecular single crystals, determined at 300 K by time-of-flight (TOF) experiments, are between 1 and 10 cm^2/V s [260]. The mobilities of organic molecular thin films are expected to be smaller due to impurities, defects, or grain boundaries that exist in the films. The weak intermolecular interaction forces in organic semiconductors, most usually van der Waals interactions with energies smaller than 10 kcal/mol (0.43 eV/molecule) [261], may be responsible for this limit, since the vibrational energy of the molecules reaches a magnitude close to that of the intermolecular bond energies at or above room temperature. In contrast, in inorganic semiconductors such as Si and Ge, the atoms are held together with very strong covalent bonds, which, for the case of Si, have energies as high as 75 kcal/mol. In inorganic semiconductors, charge carriers move as highly delocalized plane waves in wide bands and have very high mobilities. The temperature dependence of the charge carrier mobility in inorganic semiconductors follows

$$\mu = CT^{-n} \tag{6.2}$$

with $n = 3/2$. Deviations from this law are explained by a temperature-dependent mass and/or by additional scattering at optical phonons. It is important to note that mobility in inorganic semiconductors is not thermally activated, but thermally deactivated. In organic semiconductors, the temperature dependence of mobility given in relation (6.2) is not observed. Here, around room temperature, the carrier mobility is increasing as a function of temperature, that is, it is thermally activated. Only for low temperatures, a negative temperature coefficient is

Low Molecular Weight Organic Semiconductors. Thorsten U. Kampen

ISBN: 978-3-527-40653-1

observed. Therefore, the charge carrier motion is modeled as a rapid hopping between localized transport states. This hopping model takes into account the localized character of charge carriers in organic materials. The carriers polarize the surrounding lattice, and both charge carrier and its associated molecular deformation may form a quasiparticle called polaron. At the equilibrium position, the stabilization energy associated with the polaron is maximized. To move from one site to a neighboring site, the polaron must traverse a barrier.

The probability of hopping between transport states is calculated as the product of the probability of the carrier achieving energy E' and the probability of a carrier of this energy undergoing transfer to the neighbor site. In this case, the mobility is given by

$$\mu \propto T^{-n} \exp\left(-\frac{E'}{kT}\right) \tag{6.3}$$

with $n = 1$~1.5. At low temperatures, the exponential term dominates Eq. (6.3) so that, in contrast to the predictions of band theory, mobility is expected to decrease sharply. At low temperatures, band transport becomes the mechanism that takes control of carrier transport. At these temperatures, the vibrational energy is much lower than the intermolecular bonding energy and phonon scattering is very low. Thus, high mobilities are observed. At or close to room temperature, phonon scattering becomes so high that the contribution of the band mechanism to transport becomes too small and hopping begins to dominate carrier transport.

The boundary between band transport and hopping is defined by materials having mobilities between 0.1 and 1 cm^2/V s. Highly ordered organic semiconductors, such as anthracene and pentacene, have room temperature mobilities in this intermediate range, and in some cases temperature-independent mobilities have been observed.

Applying a voltage to a conductor using two contacts, the current density j

$$j = -e_0 \mu n E_x \tag{6.4}$$

can be bulk or interface determined. Here, e_0, μ, n, and E_x are the elemental charge, the carrier mobility, the carrier concentration, and the electric field, respectively. In the case of bulk-determined current densities, the metal contacts can supply freely whatever current is required. Under these circumstances, the contacts and their interfaces toward the conductor will play a negligible role in the current transport and can accordingly be ignored. The simplest case is that of a perfect insulator free of traps and with a negligible concentration of free carriers in thermal equilibrium. A continuous injection of charge carriers from the contacts can result in the formation of a space charge region in the conductor. This space charge region reduces the electrical field in the vicinity of the contacts and limits the current transport. For this case, the Poisson equation is

$$\frac{d\mathbf{E}_x}{dx} = -\frac{e_0 n}{\varepsilon \varepsilon_0} \tag{6.5}$$

where ε and ε_0 are the dielectric constant of the conductor and the permittivity of vacuum, respectively. Combining Eqs. (6.4) and (6.5) gives

$$\mathbf{E}_x \frac{d\mathbf{E}_x}{dx} = \frac{j_0}{\mu\varepsilon\varepsilon_0} \tag{6.6}$$

Assuming that the electrical field is zero at the contacts, this equation can be integrated to give

$$\mathbf{E}_x(x) = \left(\frac{2j}{\mu\varepsilon\varepsilon_0}\right)^{1/2} x^{1/2} \tag{6.7}$$

and

$$V(x) = \int_0^L \mathbf{E}_x(x)\,dx = \left(\frac{8j}{9\mu\varepsilon\varepsilon_0}\right)^{1/2} L^{3/2} \tag{6.8}$$

where L is the length of the conductor, that is, the distance between the two contacts. From Eq. (6.8), the current density results to

$$j = \left(\frac{9}{8}\right)\mu\varepsilon\varepsilon_0\left(\frac{V_0^2}{L^3}\right) \tag{6.9}$$

This equation is known as the *Mott–Gurney square law* or *Child's law* for solids, the latter in analogy with the famous *Child's law* for the vacuum diode.

The situation will change when traps exist in the solid. For the case of a single trap level in the energy at an energy E_t relative to the valence band (HOMO) (6.9) will change to [262]

$$j = \left(\frac{9}{8}\right)\mu\varepsilon\varepsilon_0\theta\left(\frac{V_0^2}{L^3}\right), \quad \text{with } \theta = \frac{N_v}{H_t}\exp\left(-\frac{E_t}{k_B T}\right) \tag{6.10}$$

where N_v is the density of states (DOSs) within $k_B T$ of the valence band edge, and H_t is the total density of traps uniformly dispersed in space.

6.1
Time-of-flight measurements (TOF)

Time-of-flight measurements determine the charge carrier mobility in thin films or crystals under an applied electrical field. [263] The carrier mobility is determined by measuring the time the carriers need to travel a certain distance, i.e. through the film or crystal. Figure 6.1 shows a typical TOF experiment. A front and back contact are prepared on the sample to apply an electrical field. Since charge carriers are excited on the front side of the sample by a short light pulse the front contact has to be semitransparent. This is achieved by either evaporating a thin semitransparent metal film, or by simply placing a metal mesh on the sample. The front contact does not have to be optimised for charge injection since charge carrier are generated within the sample and it just serves as an electrode to apply an electric field.

Charge carriers are generated by a short light pulse, where the duration of the light pulse should be much smaller than the time for travelling through the sample. The photon energy of the light should be chosen for maximum absorption, so that charge carriers are generated in a region close to the front contact.

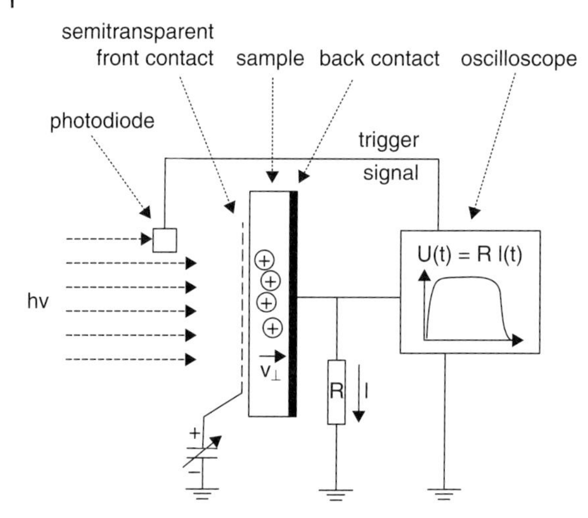

Figure 6.1 Schematic drawing of a TOF experiment. The front contact can be either a semitransparent metal film or a metal mesh. The direction of the applied electric field allows for measuring the flight time of holes.

This region should be much smaller than the thickness of the sample L. The charge carriers are separated in the electric field and depending on the polarity of the applied field either positive or negative charge carriers will travel through the sample towards the back contact. In a very simple picture the volume of charge carriers may be viewed as a thin disk moving from the front contact towards the back contact. For a constant homogenous field across the sample the charge carriers will assume a constant average drift velocity, $v_\perp$. Hence a constant current $I(t)$ is measured as a voltage drop $U(t)$ across a resistor R using an oscilloscope. The data acquisition is started by a trigger signal which is provided by a photo diode. The current $I(t)$ lasts until the carrier disk reaches the back electrode at time τ; that is, a quasi-rectangular pulse should be detected. The mobility $\mu_\perp$ is given by

$$\mu_\perp = \frac{v_\perp}{E} = \frac{L^2}{\tau U} \tag{6.11}$$

where U is the voltage applied to the electrodes and $\mu_\perp$ is the mobility along the given direction of E.

Figure 6.2 displays TOF signals for different polarities of the applied electric field. The top curve shows a positive pulse originating from holes, while the bottom curve is due to electrons measured under reversed field direction. The area under the curves scales with the number of charge carriers and τ is determined from the inflection point of the trailing edge of the pulse.

Samples investigated by TOF have a typical thickness of 0.1–1 mm and voltages of several hundred Volts are applied. TOF measurements are limited by a typical time resolution in the data acquisition of about 1 ns. This restricts TOF measure-

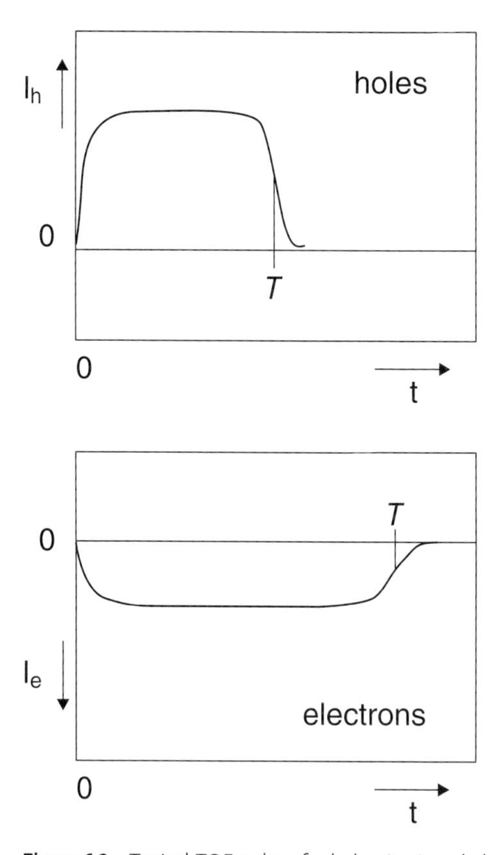

Figure 6.2 Typical TOF pulses for holes (top) and electrons (bottom),

ments to a certain minimal quotient $L/\mu_{\perp}$. Hence a high charge carrier mobility in a very thin sample may not be measurable with this technique. In addition the sample should have a significantly low dark conductivity, that is, dark current, so that the TOF signal can be well resolved.

TOF measurements were employed by N. Karl and co-workers [i] to investigate the charge carrier mobility in thin films of PTCDA. They prepared PTCDA films with a thickness ranging from 280–430 nm on different substrates like Si, SiO_2, ITO, and Au films on fused silica. The conditions for the PTCDA vapor deposition have been varied, for example, thin films have deposited in HV and UHV, respectively. Therefore, PTCDA films with a wide range of quality have been produced, which is evident from the variation of the width of the symmetric (102) Bragg reflection in the X-ray rocking curves. Here, the rocking width is a measure of order and alignment of the PTCDA film. The highest quality films with lowest rocking width have been prepared on Si-surfaces under UHV conditions. The lowest quality films, on the other hand, are found on ITO and Au/fused silica substrates. In a next step the mobility of electrons at room temperature is determined by TOF measurements. In preparation for these measurements semitrans-

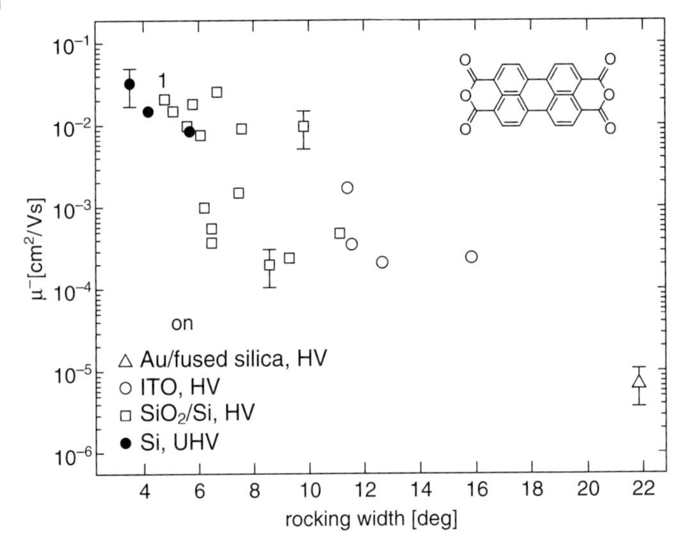

Figure 6.3 TOF electron mobilities measured in vacuum vapor-deposited PTCDA thin films grown on different substrates. Disorder increases with increasing rocking width [from Reference i].

parent gold films were deposited onto the PTCDA films. Figure 6.3 shows the electron mobility in different PTCDA films as a function of the respective rocking width. The electron mobility is found to decrease with increasing rocking width indicating that the highest mobility is found in the highest crystalline quality films. There seems to be a general trend, that films grown on Si or SiO_2 have better quality than those grown on ITO or Au. But even for films grown on SiO_2 a large variation in the electron mobility by two orders of magnitude is found. The highest mobility of $\mu = 3 \times 10^{-2}\,cm^2/V$ is found in PTCDA films grown on Si.

6.2 Thin Field Effect Transistor (TFT) Mobilities

While TOF spectroscopy measures the mobility of charge carriers perpendicular to the film, the field effect transistor geometries shown in Figure 6.4 offer the possibility to measure the carrier mobility parallel to the film. The important features of a field effect transistor are the electrical contacts called source (S) and drain (D), and the electrode called gate (G) which is electrically isolated from the source and the drain. In a metal-oxide-semiconductor field effect transistor (MOSFET) the gate is separated from the semiconductor substrate by a layer of insulating oxide.

The inset of Figure 6.4 shows a MOSFET structure with source, drain, and gate on top of the semiconductor substrate. This geometry is called top gate geometry and is often used for inorganic semiconductors. The gate oxide prevents any current flow when a gate voltage U_G is applied to the gate. Instead the gate voltage

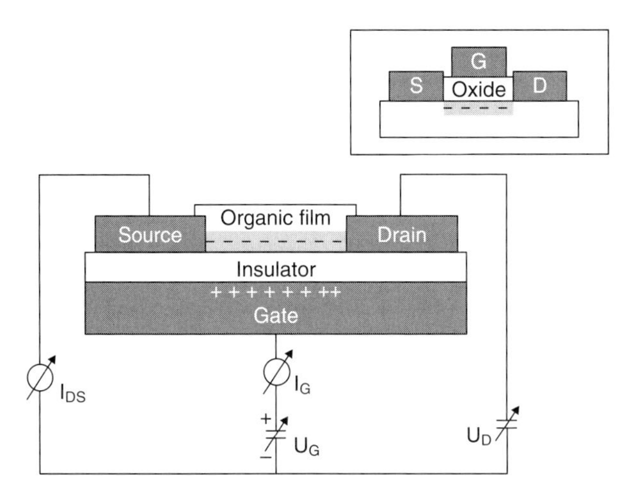

Figure 6.4 Bottom gate field effect transistor geometry for charge carrier mobility measurements in organic thin films. Inset shows top gate geometry for comparison.

results in an electric field that modulates the carrier concentration in the semiconductor region close to the semiconductor-oxide interface. This effect is called field effect. Applying a positive bias to the gate will increase the concentration of electrons in this region. Hence, a n-type channel is formed between source and drain. For measurements on thin films of organic semiconductors the bottom gate geometry is used because it offers the opportunity to deposit the organic thin film in the last preparation step.

The charge carrier mobility can be either determined from the transconductance or the saturation current. For a drain source voltage U_{DS} below the saturation limit, that is, in the linear part of the $I_{DS}(U_{DS})$ characteristics, the transconductance is given by

$$\left(\frac{\partial I_{DS}}{\partial U_G}\right)_{U_{DS}=\text{const.}} = \pm\frac{Z}{X}\mu C_i U_{DS}, \tag{6.12}$$

where X and Z channel length and width, respectively, and C_i is the capacitance per unit area between the gate electrode and the organic film. The plus and minus sign are for *n*- and *p*-type conduction, respectively.

In the saturation limit of the $I_{DS}(U_{DS})$ characteristics the saturation current $I_{DS,sat}$ is given by

$$I_{DS,sat} = \pm\frac{Z}{X}\mu C_i (U_G - U_0)^2, \tag{6.13}$$

with the threshold voltage U_0 taking into account the field free carrier concentration as well as details of the internal charge distribution at the interface and at the contacts.

Figure 6.5 shows the $I_{DS}(U_{DS})$ characteristics of a TFT with a 53 nm thick palladiumphthalocyanine (PdPc) film [i]. The films were vapour-deposited onto grown a 225 nm thick SiO_2/Si_3N_4 insulator layer at an elevated substrate temperature of

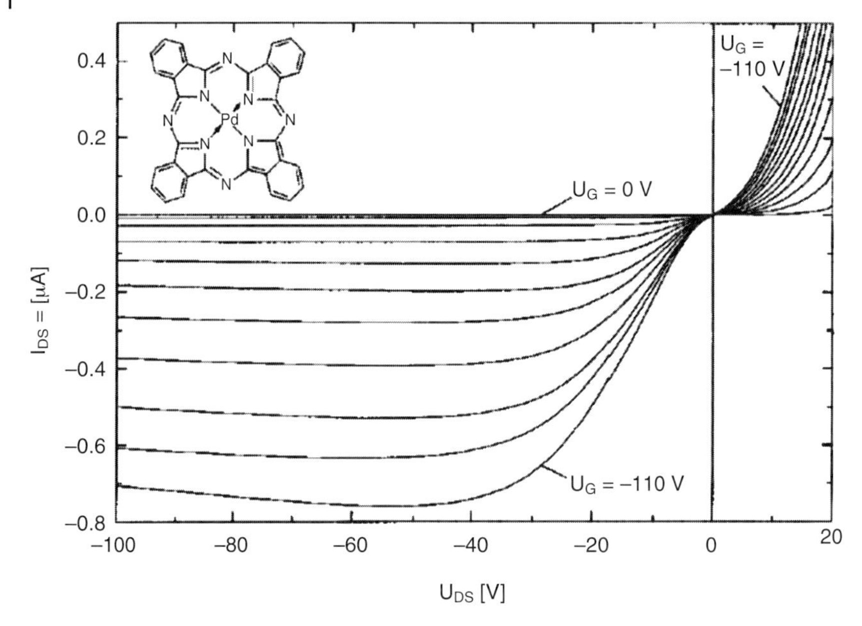

Figure 6.5 Organic thin film field effect transistor with palladiumphthalocyanine (PdPc). $I_{DS}(U_{DS})$ characteristics have been taken with gate voltages ranging from 0 to −110V in 10V steps [from Reference 9].

150 °C. Channel length X and width Z are 15 μm and 2 mm, respectively. The $I_{DS}(U_{DS})$ characteristics show the typical behavior of a field effect transistor. The current I_{DS} saturates for $U_{DS} > 40$ V and it's negative sign leads to the conclusion that a *p*-type channel is formed, that is, charge transport carriers are holes. With Eq. (6.13) the hole mobility amounts to 2.5×10^{-5} cm^2/Vs.

6.3
I/*V* Curves of Ag/GaAs(100) Schottky Contacts

Forrest *et al.* have investigated the properties of organic modified Schottky contacts on elemental and compound semiconductors. Indium contacts on p-Si usually result in contacts showing nearly ohmic *IV* characteristics. Depositing a PTCDA layer of 200 nm prior to the formation of the In contact results in rectifying contacts. The barrier heights for In/PTCDA/p-Si Schottky barriers are found to be (0.74 ± 0.02) eV, which is considerably larger than the barrier height of (0.55 ± 0.02) eV of Ag/p-Si reference diodes. On n-Si, the barrier heights of modified In contacts are found to be (0.61 ± 0.01) eV. This is lower than the barrier height of 0.76 eV found for Au/n-Si reference diodes. For both types of substrate, the ideality factors are found to be larger for the organic modified Schottky contacts. Holes are found to be the dominant charge carrier type in PTCDA. The mobilities as high as

1.4 cm^2/V s are obtained from the part of the IV characteristics, which is determined by space charge, limited currents. High mobilities are found in films deposited with high deposition rates (3 nm/s), which is assumed to result in superior crystalline quality of the PTCDA films. In addition, PTCDA and DiMe-PTCDI were used to modify metal/GaAs Schottky contacts. Their IV characteristics can be understood using the thermionic-emission-space-charge-limited transport model. Again, the organic modification using PTCDA leads to higher and lower barrier on p- and n-type doped substrates, respectively, compared to the reference diodes. Using DiMe-PTCDI instead of PTCDA results in high barrier heights of 0.85 eV on n- and p-type doped substrates.

The results presented in this chapter show that larger barrier heights result from the exposure of the organic modified Schottky contacts to air. Performing sample preparation and characterization in UHV reveals that the barrier heights can be decreased as a function of organic layer thickness. The major mechanism responsible for the decrease in barrier height is an increase in image force lowering at the metal–semiconductor interface induced by the organic film [264].

The IV characteristics of Ag contacts on both types of substrates as a function of d_{PTCDA} are shown in Figure 6.6 as solid lines. The curves of the two unmodified Ag contacts may be described by thermionic emission of carriers over a barrier, that is, by (see, for example, Reference [265])

$$I = FA^{**}T^2 \exp\left(\frac{-\varphi_{\mathrm{B}n}}{k_{\mathrm{B}}T}\right)\exp\left(\frac{e_0 V_a}{nk_{\mathrm{B}}T}\right)\left[1-\exp\left(\frac{e_0 V_a}{k_{\mathrm{B}}T}\right)\right] \tag{6.14}$$

As usual, A^{**}, k_B, e_0, F, and T are the effective Richardson constant of n-type GaAs, Boltzmann's constant, elementary charge, area of the contact, and temperature, respectively, and I is the current. The ideality factor accounts for the dependence of the barrier height φ_{Bn} on the applied voltage V_a. For applied voltages $V_a > 3k_BT/e_0 \approx 0.08$ eV at RT, the experimental IV characteristics exhibit a linear behavior in a semilogarithmic plot. Least square fits to these parts of the data give zero-bias barrier heights φ_{Bn} of (0.82 ± 0.01) eV and (0.59 ± 0.01) eV and ideality factors n of 1.1 ± 0.01 and 1.09 ± 0.01 for the Ag/GaAs(100) and Ag/S–GaAs(100) contacts, respectively. It is important to note that the thermionic emission current depends linearly on the contact area, but exponentially on the barrier height. A change of 50 meV in barrier height changes the current by one order of magnitude. In this sense, Schottky contacts are very sensitive to lateral inhomogeneous distributions of the barrier height, especially if the inhomogeneities exhibit lower barrier heights.

The voltage drop

$$V_R = RI \tag{6.15}$$

across a series resistance R with contribution from the GaAs bulk, its ohmic back-contacts, and external connections result in a deviation from straight lines for a high forward biases.

Figure 6.6a shows the IV characteristics of Ag/PTCDA/GaAs(100) for different PTCDA layer thickness (d_{PTCDA}). With increasing d_{PTCDA}, the current is increasing, indicating that the barrier height of the contact is effectively lowered.

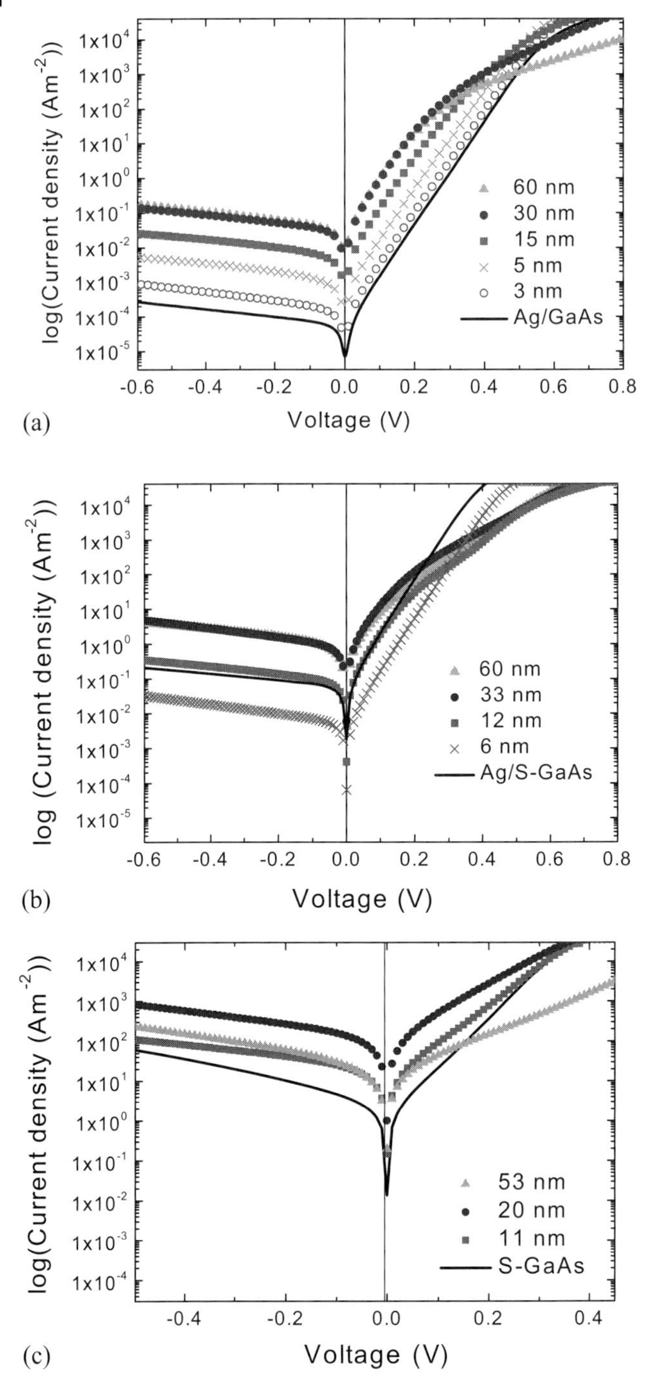

Figure 6.6 *IV* characteristics of (a) Ag/PTCDA/GaAs(100), (b) Ag/PTCDA/S–GaAs(100), and (c) Ag/DiMe-PTCDI/S–GaAs(100) Schottky contacts as a function of the organic layer thickness at room temperature. The contact area is $2.1 \times 10^{-7}\,m^2$.

For reverse biases and forward biases below 0.2 V, a behavior typical for thermionic emission is observed. For a $d_{PTCDA} > 5\,nm$ and applied voltages larger than 0.2 V, the *IV* characteristics are now influenced by the transport properties of the organic material. Transport through organic materials is governed by space charge limited currents (SCLC), which can be described by the Mott–Gurney law (6.9).

The *IV* characteristics of the Ag/PTCDA/S–GaAs(100) contacts are displayed in Figure 6.6b. Opposite to the behavior of the Ag/PTCDA/GaAs(100) contacts, the current first decreases as a function of d_{PTCDA}, while for $d_{PTCDA} \geq 12\,nm$ the current becomes larger compared to the unmodified contact in the reverse and the low forward bias region.

IV characteristics for various DiMe-PTCDI interlayer thicknesses in Ag/S–GaAs diodes are displayed in Figure 6.6c. Clearly the current is larger for all diodes that contain DiMe-PTCDI interlayers than for the one without. Therefore, the effective barrier height is lowered for all interlayer thicknesses.

The *CV* characteristics of the organic modified contacts as a function of d_{PTCDA} are shown in Figure 6.7. The flat band barrier height for the unmodified Ag/GaAs(100) contact is (0.96 ± 0.01) eV. The zero-bias barrier height is lower than the flat-band barrier height due to the image force lowering at the metal–semiconductor interface. For $d_{PTCDA} \leq 30\,nm$, the *CV* characteristics of organic modified contacts are comparable to the one of the unmodified contact. In this thickness region, the capacitance of the organic film is much larger than the capacitance of the space charge region of the GaAs and does not contribute to the overall capacitance. Therefore, *CV* characteristics of the Ag/PTCDA/GaAs(100) contact are completely determined by the capacitance of the space-charge region in the inorganic semiconductor, which does not change upon deposition of PTCDA. For $d_{PTCDA} > 30\,nm$, SCLC starts dominating the *IV* characteristics.

For $d_{PTCDA} < 30\,nm$, the flat-band barrier heights determined from the *CV* characteristics of organic modified contacts on both types of substrates are of the same

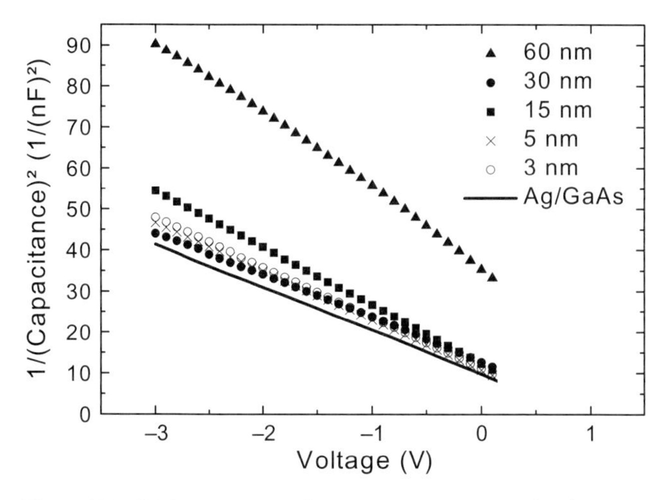

Figure 6.7 *CV* characteristics of Ag/PTCDA/GaAs(100) Schottky contacts as a function of the d_{PTCDA} at room temperature. The contact area is $2.1 \times 10^{-7}\,m^2$.

value as for the unmodified contacts. The *CV* characteristics are completely determined by the capacitance of the space-charge region in the inorganic semiconductor, which does not change upon deposition of PTCDA. For $d_{PTCDA} > 30\,nm$, SCLC starts dominating the *IV* characteristics and the capacitance of the organic layer starts to contribute significantly to the *CV* characteristics.

For the determination of the barrier heights for the organic modified Ag/GaAs(100) Schottky contacts, the contribution of SCLCs and series resistances to the charge transport have to be taken into account in addition to thermionic emission. The voltage drop over each contribution adds up to the total applied voltage:

$$V = V_a + V_o + V_r \tag{6.16}$$

Inserting (6.14) and (6.15), and $V_r = RI$ in (6.16), the experimental *IV* curves can be fitted, using φ_{Bn}, n, μ, and R as fitting parameters. For the Ag/DiMe-PTCDI contacts, the barriers were derived from the saturation current extrapolated to zero bias employing thermionic emission theory.

Figure 6.8 shows the resulting effective barrier heights for the three types of contacts as a function of d_{PTCDA}. The difference in barrier heights of the Ag/GaAs(100) and Ag/S–GaAs(100) contacts is attributed to a sulfur-induced interface dipole as has been discussed in Section 4.2.

The change in barrier height as a function of d_{PTCDA} can be quantitatively explained using the energy level diagrams determined by photoemission spectroscopy shown in Figure 6.9 [266]. In both cases, PTCDA does not change the electronic properties of the As-prepared GaAs substrate, that is, the band bending is constant. For the plasma-treated substrate, the band bending in GaAs is determined to be (0.8 ± 0.01) eV, which is same as in Ag/GaAs(100) contacts. Starting with the lowest d_{PTCDA}, the effective barrier height is first constant and then decreases, indicating that $E_{LUMO,trans}$

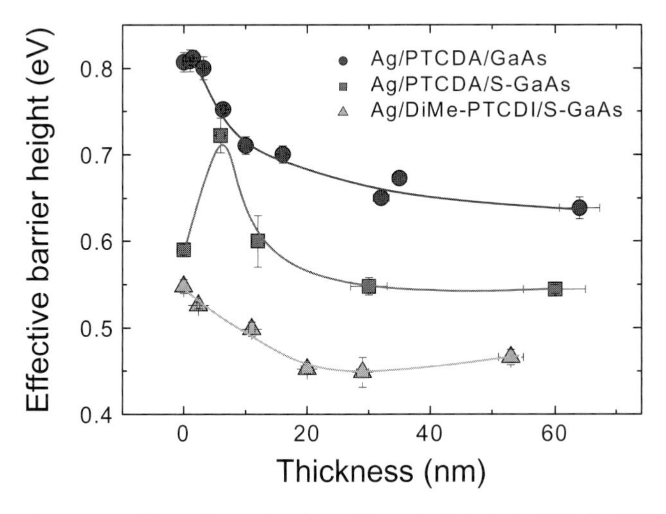

Figure 6.8 Effective barrier heights of Ag/PTCDA/S–GaAs(100), Ag/PTCDA/GaAs(100), and Ag/DiMe-PTCDI/S–GaAs(100) Schottky contacts as a function of organic layer thickness.

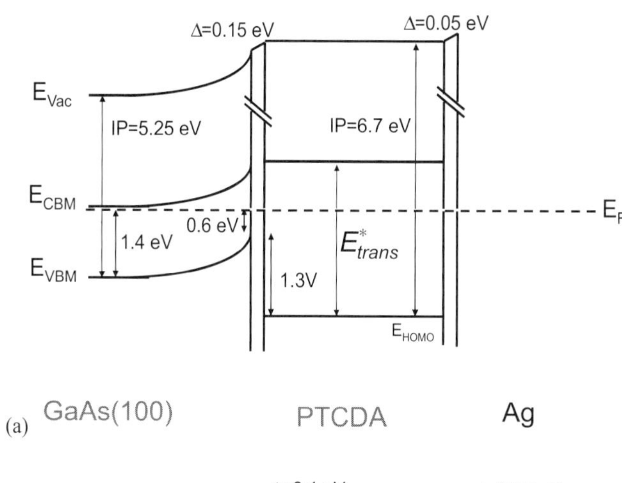

(a)

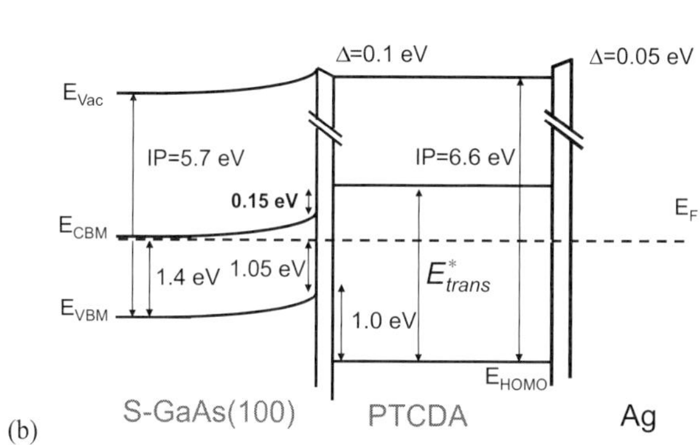

(b)

Figure 6.9 Band diagrams of (a) Ag/PTCDA/GaAs(100) and (b) Ag/PTCDA/S–GaAs(100) interfaces.

is approximately at the same energy or below the CBM of the GaAs at the interface. For Ag/PTCDA/S–GaAs contacts, the barrier height initially increases for low d_{PTCDA} indicating that $E_{\text{LUMO,trans}}$ is now above the conduction band maximum. The distance $E_{\text{LUMO,trans}}$ −conduction band maximum can be estimated by the initial increase in the barrier height to be 0.15 eV. With these results, the distance between $E_{\text{LUMO,trans}}$ and the high-energy edge of the HOMO can be estimated to be 2.55–2.8 eV. A further increase in d_{PTCDA} results in a decrease in effective barrier height. The decrease in effective barrier height observed for both types of substrate can be explained by a strong contribution of the low dielectric constant material PTCDA ($\varepsilon = 2$) to image force lowering at the PTCDA–GaAs interface. The image force lowering can be calculated using (see Reference [18])

$$\Delta\varphi_{\text{if}} = e_0 \left[\frac{2e_0^2 N_D (\varphi_{\text{Bn}} - \xi - e_0 V_a - k_B T)}{(4\pi)^2 (\varepsilon_b \varepsilon_0)^3} \right]^{1/4} \tag{6.17}$$

Here, $\xi = 20\,\text{meV}$ is the position of the Fermi level with respect to the CBM in the bulk, ε_b is the dielectric constant of the inorganic semiconductor, and ε_0 is the permittivity in vacuum. For Ag/GaAs(100) contacts, image force lowering amounts to 50 meV. In a very simple approach, one may substitute ε_b by the dielectric constant of PTCDA. This results in an image force lowering of about 200 meV, which is in good agreement with the experimentally determined decrease in barrier height of 150 meV.

Finally, the evolution of effective barrier height for DiMe-PTCDI interlayer will be discussed. Compared to Ag/PTCDA/S–GaAs(100) contacts, the barrier height already drops for the lowest organic film thickness, indicating that the LUMO should be at or below the conduction band maximum of GaAs. This seems to be in contradiction with the energy level alignment obtained from the photoemission spectra. There, the LUMO was found to be 0.11 eV above the CBM, which should result in an initial increase in barrier heights. On the other hand, the AFM picture in Figure 3.7 clearly shows that the S-passivated GaAs(100) is not completely covered by the DiMe-PTCDI. This results in an inhomogeneous distribution of barrier heights at the interface of Ag/DiMe-PTCDI/S–GaAs(100) contacts. For low DiMe-PTCDI coverages, the barrier height will be lower in areas, which are not covered by DiMe-PTCDI. Therefore, the current will predominantly flow through these regions. With increasing DiMe-PTCDI, the image force lowering becomes more and more important. Now, the barrier height is lower in the regions covered by DiMe-PTCDI and current will flow through these regions. The situation is different for PTCDA. During the growth of PTCDA, a monolayer is formed followed by island growth. These islands, as can be seen in Figure 3.7, cover the substrate in a closed packed fashion. In other words, the PTCDA completely covers the GaAs(100) surface. This results in a lateral homogenous distribution of the barrier height, and the initial increase in barrier height due to the LUMO being above the CBM in GaAs can be observed.

Due to the image force lowering, the barrier heights at Ag/DiMe-PTCDI/S–GaAs(100) contacts show a minimum barrier of 0.45 eV that is reached for a DiMe-PTCDI layer thickness of 30 nm. For higher thicknesses, the barrier height increase slightly. The increase in barrier height for the highest coverage can be attributed to the different molecular ordering in PTCDA and DiMe-PTCDI films. While PTCDA molecules are lying flat on the substrate surface, the molecular plane of DiMe-PTCDI molecules is tilted in the range of 56° with respect to the substrate plane. Assuming that DiMe-PTCDI shows the same anisotropy in carrier mobility as PTCDA, a tilt of the molecular planes would result in a lower mobility for the charge carriers traveling through the organic film in a direction perpendicular to the substrate surface. As a consequence, SCLCs become important at lower organic film thickness and result in lower current densities, which may be interpreted as a higher barrier height.

6.4 Charge Carrier Mobilities

Another parameter determined from the evaluation of the *IV* characteristics of the PTCDA-modified GaAs(100) Schottky contacts using the thermionic-emission-

Table 6.1 Charge carrier mobilities in PTCDA films grown on nonpassivated and passivated GaAs(100) substrates as a function of film thickness.

PTCDA thickness (nm)	Ag/PTCDA/GaAs(100) (μ_{eff}/cm^2/V s)	Ag/PTCDA/S–GaAs(100) (μ_{eff}/cm^2/V s)
0		
3	$(3.0 \pm 1.5) \times 10^{-2}$	
5	$(5.7 \pm 0.7) \times 10^{-3}$	
6		$(3.2 \pm 0.9) \times 10^{-2}$
12	$(1.8 \pm 0.02) \times 10^{-3}$	$(3.3 \pm 1.0) \times 10^{-3}$
15	$(3.0 \pm 0.04) \times 10^{-3}$	
30	$(1.3 \pm 0.04) \times 10^{-2}$	$(2.2 \pm 0.4) \times 10^{-2}$
60	$(4.2 \pm 0.2) \times 10^{-3}$	$(4.2 \pm 0.6) \times 10^{-2}$

space-charge-limited transport model is the charge carrier mobility μ_{eff} in the organic materials. The different values of μ_{eff} in the two types of contacts imply the difference in the quality of the PTCDA layers grown on different GaAs(100) surfaces. As can be seen in Table 6.1, μ_{eff} is generally higher for PTCDA grown on the S-passivated GaAs(100) indicating a higher crystallinity. This agrees with a previous LEED study of PTCDA grown on different GaAs(100) surfaces, where the passivated GaAs(100) surface results in a better-ordered PTCDA film.

The mobilities presented in Table 6.1 compare quite well to mobilities presented by Hudej *et al.* and Marktanner. Hudej *et al.* investigated the transient photoresponse in Au/PTCDA/In structures [267]. The mobilities of electrons perpendicular to the molecular layers, that is, the (102) planes, are found to increase with the applied electric field and saturates for fields higher than 5×10^4 V/cm. The maximum mobility is 3×10^{-2} cm^2/V s. Similar mobilities have been determined by Marktanner in PTCDA films prepared under optimized conditions on Si in UHV [268]. The sign of the TOF signal clearly indicates that the charge is carried by electrons.

6.5 Simulation of *IV* Curves

A more detailed study of the image force lowering in organic-modified Schottky contacts has been performed by Aldo *et al.* [269]. They have calculated the conduction band edge as a function of the PTCDA thickness considering the three different dielectric materials contained in the metal/PTCDA/GaAs structure. In addition, the image force in PTCDA and GaAs is taken into account. For further details, see Reference [269]. The conduction band edge profile of the Schottky diode for a constant electric field of 200 kV/cm, which is a typical value in such devices, is shown in Figure 6.10. Neglecting the singularity at the interface [270] and only

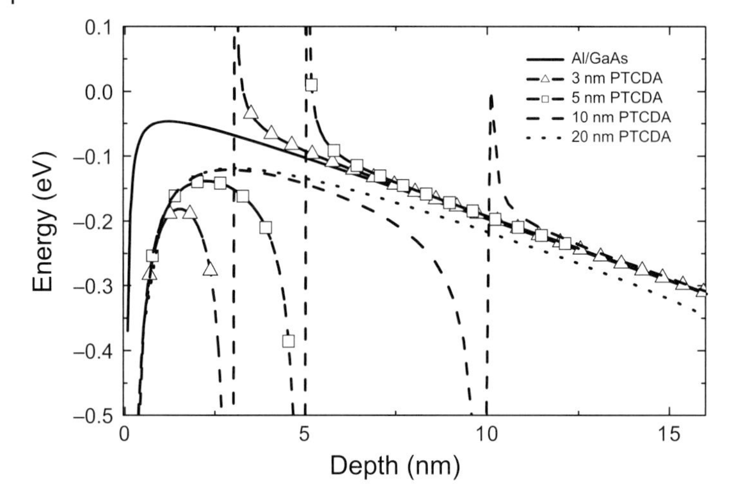

Figure 6.10 Conduction band edge for the PTCDA/GaAs Schottky diode as a function of the PTCDA thickness. The calculations are performed with a constant electric field of 200 kV/cm.

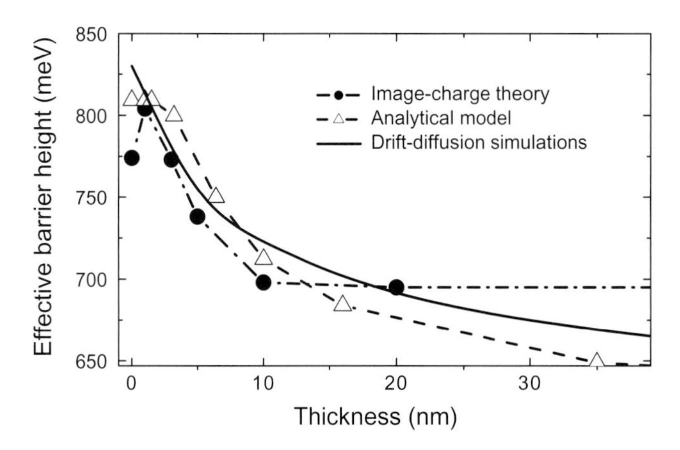

Figure 6.11 Effective barrier heights of Ag/PTCDA/GaAs(100) Schottky contacts as a function of d_{PTCDA} at room temperature. The three curves refer to the barrier height obtained with the image-charge theory (full circle), the barrier height extracted from experiments by using the analytical model of (open triangle), and the barrier height extracted from experiment by using the drift-diffusion simulations (solid line).

considering the maximum value of the conduction band edge as an estimation of the barrier height, a barrier reduction of 110 meV is found. Moreover, for PTCDA thicknesses larger than 10 nm, the barrier reduction is negligible.

In Figure 6.11, the barrier height obtained with this method is shown as a function of the PTCDA thickness and compared with the barrier heights extracted from the experimental data presented in Section 6.3. There is an overall good agreement between the two curves.

In a second approach, the *IV* characteristics of organic-modified Schottky contacts were simulated using a two-dimensional drift-diffusion simulator. Such approach has already been used to describe the interplay between contact barrier height and mobility on the output characteristics of organic thin film transistors [271].

The charge transport in an organic semiconductor can be obtained by a proper definition of the following quantities: mobility, density of the states, equivalent doping, trap distribution, and band alignment at organic–inorganic semiconductor and organic–metal interfaces.

Based on Monte Carlo simulations [272], the field dependence of the mobility is given by

$$\mu(E) = \mu_0 \cdot \exp\left(\sqrt{\frac{E}{E_0}}\right) \tag{6.18}$$

where $\mu_0 = 10^{-3}\,\mathrm{cm^2/V\,s}$ is the low field mobility and $E_0 = 10^5\,\mathrm{V/cm}$ is the critical field [273, 274].

DOS in organic materials is different from that of inorganic semiconductors. However, we can define an effective DOS as a single equivalent level on the edge of the bands. For the temperature considered here, all states are thermally accessible and we can set the effective density of the states equal to the density of molecules. The GaAs has a thickness of 475 μm with a doping concentration of $2 \times 10^{17}\,\mathrm{cm^{-3}}$. For PTCDA, according to the photoemission spectroscopy measurements, we use a transport gap of 2.55 eV, an electronic affinity $\chi = 4.15\,\mathrm{eV}$, and a DOS $N = 2 \times 10^{22}\,\mathrm{cm^{-3}}$. Further, the band alignment in metal/PTCDA/GaAs shown in Figure 6.9 was used. A contact resistance of 300 Ω has been considered for the GaAs back contact.

The simulated output currents are shown in Figure 6.12 for several PTCDA layer thicknesses. Here, the only fitting parameter is the barrier height. The *IV* characteristics present two different regimes. For biases below 0.3 V, the typical behavior of Schottky diode is obtained, while for biases larger than 0.3 V the transport is governed by SCLCs. For even larger biases the current is limited by the contact resistance. The barrier heights obtained from the simulation are also presented in Figure 6.11. The overall agreement between the simulation and the experimental results is remarkably good.

6.6 Chemical Stability

Ag/PTCDA/GaAs(100) contacts of a sample with $d_{\mathrm{PTCDA}} = 30\,\mathrm{nm}$ were subsequently exposed to N_2, H_2, and O_2. Figure 6.13 shows the *IV* and *CV* characteristics of this contact taken under UHV conditions after each exposure. The *IV* and *CV* characteristics taken after preparing the sample, after storing it for 5 days in UHV,

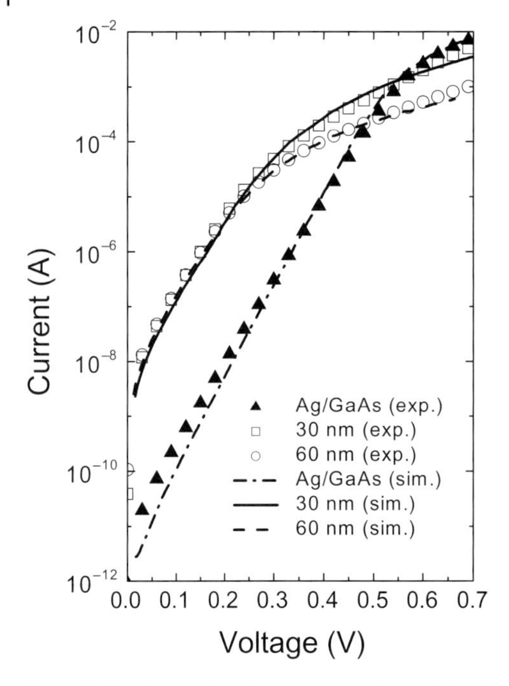

Figure 6.12 Comparison between experimental (exp.) and simulated (sim.) *IV* curves of the PTCDA/GaAs Schottky diode for several values of the PTCDA layer thickness.

and after an exposure of 1 bar H_2 for 15 min are almost identical. A deviation occurs for high forward bias, where the ohmic series resistance starts to contribute to the *IV* characteristics. This is due to the fact that gas exposures and electrical characterizations have been performed in different UHV chambers connected by a gate valve. Therefore, the electrical connection to the Ag contacts had to be applied several times by a gold tip, which may result in a small variation of the contact resistance.

After an exposure to 1 bar N_2 for 15 min, the current decreases in the low-voltage region indicating that the effective barrier height is enhanced. The decrease in current is even more obvious after an exposure to 1 bar O_2 for 15 min. Here, the current in reverse direction is the same as for the bare Ag/GaAs(100) contact indicating that the barrier heights for both contacts are approximately the same. The slope of the *IV* characteristics at forward biases deviates significantly from that of the bare Ag/GaAs(100) contact, which can be attributed to a larger contribution of the organic film to the *IV* characteristics due to a decreased carrier mobility, or, in other words, an increased resistance. The *CV* characteristics after 5 days storage in UHV, after H_2 and N_2 exposure, are comparable to the one of the bare Ag/GaAs(100) contacts and result in the same flat-band barrier height. After O_2 exposure, the overall capacitance decreases. The changes in *IV* and *CV* characteristics after O_2 exposure can be explained by the formation of deep traps due to the O_2

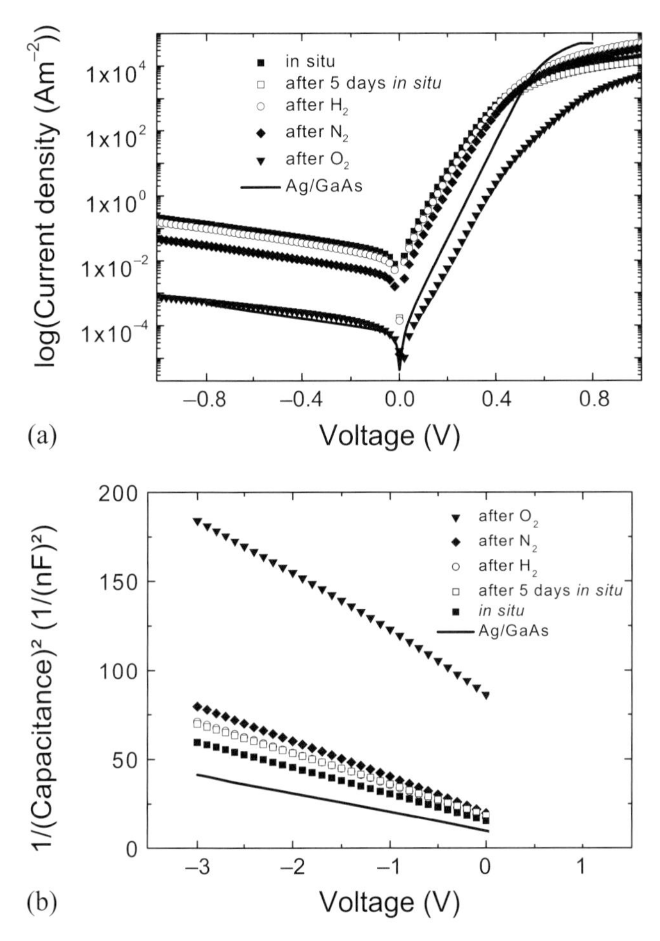

Figure 6.13 *IV*- and *CV* characteristics of a Ag/PTCDA(30 nm)/GaAs(100) after subsequent exposure to H_2, N_2, and O_2. Solid lines show *IV*- and *CV* characteristic of an Ag/GaAs reference diode.

resulting in a reduced electron concentration. The formation of traps is confirmed by deep-level transient spectroscopy (DLTS) measurements [275] (Section 6.7).

The Fermi level within the PTCDA is shifted by 160 meV from its original position down toward the middle of the band gap. This results in a larger energy discontinuity between the CBM of the GaAs and the LUMO of the PTCDA increasing the effective barrier height and counterbalancing the image force lowering. The dash-dotted lines in Figure 6.13 schematically show this. Such shifts in the HOMO toward the Fermi level upon oxygen exposure have been observed on 5,10,5,15,20-zinc-tetraphenyl porphyrin (ZnTPP) films grown on Mg, Al, Ag, and Au substrates [200]. After an exposure to oxygen of 4 Torr for 5 min, the Fermi level is found to shift toward the HOMO in films grown on Mg and Ag. The energy distance between HOMO and Fermi level is even increased for films grown on

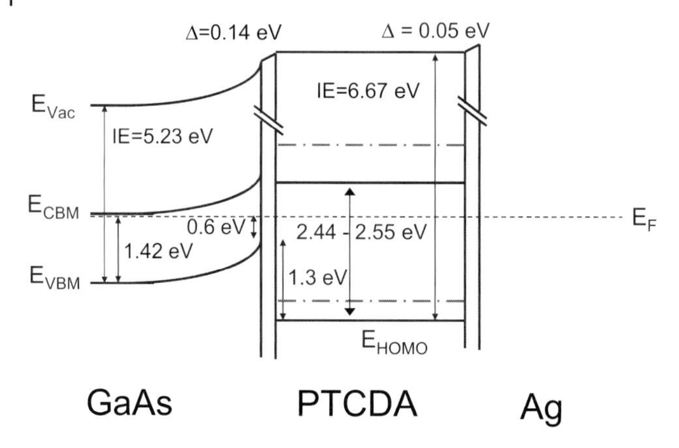

Figure 6.14 Band diagram of Ag/PTCDA/GaAs(100) interface before and after (dasheddotted lines) exposure to O_2.

Al, while films grown on Au show no shift at all. The direction and magnitude depends on the metal. Also CuPc films show a shift of the Fermi level toward the HOMO by about 0.1 eV upon exposure to 10^8 L O_2 [276, 277]. Moreover, changes in the density of filled electronic states in the bad gap below the Fermi level were observed and attributed to oxygen-induced electronic gap states. The original position of the Fermi level on freshly prepared samples could be recovered after exposure to 10^8 L H_2 or annealing in UHV at 380 K. In principal, two effects have to be taken into account: oxygen-induced changes in the organic films and oxidation of the metal at the interface. In some cases, oxide layers have been found to improve the performance of OLEDs [278, 279].

Recent internal photoemission spectroscopy investigations performed in air seem to indicate that the properties of organic/GaAs(100) interface are not modified by the exposure to air [280]. Therefore, the increase in barrier height is due to the contamination of the PTCDA/Ag interfaces resulting in interface states. This interpretation is supported by *ex situ* impedance investigations on Ag/PTCDA/Ag samples, the details of which are given elsewhere [281]. From the impedance measurements, a Gaussian distribution of density of interface states with a width of 0.73 eV is determined. Moreover, at an applied dc voltage of 0.25 V the Fermi level is crossing the "center" of the DOS. Further *in situ* and *ex situ* investigation using internal photoemission spectroscopy and impedance spectroscopy will be performed to investigate the impact of oxygen on the electronic properties of organic–inorganic semiconductor and organic–metal interfaces.

6.7
Deep level transient spectroscopy

One advantage of inorganic semiconductors is, that their electrical properties can be changed over a wide range by incorporating small amounts of impurities or defects. Doping, to name an example, is the intentional incorporation of donor- or acceptor-type impurities with energy levels just below the conduction band minimum and above the valence band maximum, respectively. Hence, these impurities are also called shallow impurities or shallow levels. While doping concentrations amount to about one impurity per one million host atoms, their contribution to the charge carrier concentration is by orders of magnitude larger than the intrinsic carrier concentration.

Deep levels, on the other hand, are occupied or unoccupied states within the band gap of the semiconductor. Contrary to doping levels, their energy position is further away from the band edges and closer to the centre of the band gap. Due to the relatively large energy distance to the respective band edges of about 0.5–1 eV, electrons or holes within these deep levels cannot be thermally excited at RT into their respective band. Hence, electrons or holes within those deep levels do not longer participate in charge transport, i.e. they are trapped. In addition of decreasing the lifetime of charge carriers, deep levels also serve as non-radiative recombination centres for electron-hole pairs.

Like in inorganic semiconductors deep levels also deteriorate the properties of organic semiconductors and devices like OLEDS. [282, 283]. Deep levels are caused by broken bonds, strain associated with displacement of atoms, and differences in electronegativity or core potentials between impurity and host.

Another mechanism which may influence device performance are dipoles and their reorientation in electrical fields. Dipoles are the result of a net charge transfer due to next neighbor interaction. They appear in interatomic bonds with ionic character, at interfaces, or in molecules. Reorientation of dipoles in OLEDs seems to play an important role in recovery of degradation [284]. The dipole induced internal electric field under the reverse bias seems to lead to the recovery of the short-term degradation of OLEDs, normally working at forward biases.

Deep levels and dipoles can be identified by charge deep-level transient spectroscopy (*Q*-DLTS) [285] and the feedback charge capacitance-voltage method (FCM) [286]. I. Thurzo and co-workers have used these techniques to investigate organic heterostructures [287]. In both techniques an alternating voltage is applied to the sample, and the transient charge is measured as a function of time. The applied voltage consists of a bias U_g and a pulse ΔU superimposed on it. The idea is that U_g shifts the Fermi level in such a way across the band gap that the deep levels are left charged or uncharged depending on the character and their relative energy position within the band gap. This state is changed by ΔU and the resulting flow of charge is measured as a function of time. *Q*-DLTS measurements are either performed at constant temperature measuring the transient charge Q(t), or by varying the temperature and measuring the change in charge as a function of the temperature.

Dipoles, on the other hand, may couple to the alternating electric field changing the overall capacitance of the sample. Deep levels or polarization effects cause the charge transient signal $Q(t)$ to be dependent or independent on the bias voltage, respectively.

The biasing and sampling sequences in *Q*-DLTS feedback charge C-V are shown in Figure 6.15. Instead of measuring $Q(t)$ quasi-continously, the charge is meas-

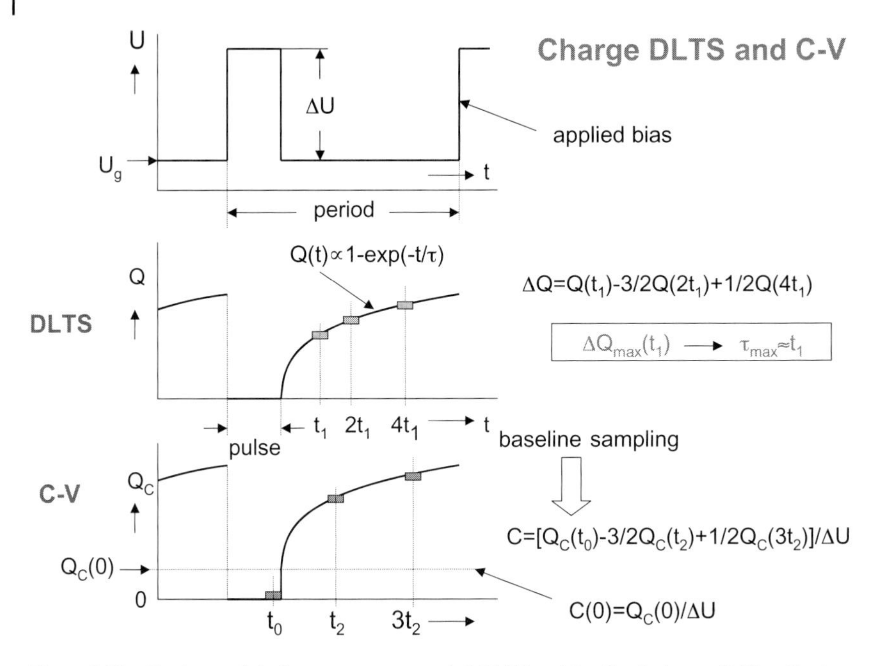

Figure 6.15 Biasing and timing sequences used *Q*-DLTS and feedback-charge *C–V* methode [from Reference 289].

ured at three defined times using a three channel correlator. In *Q*-DLTS measurements are done at t_1, $2t_1$, and $4t_1$ relative to the trailing edge of the pulse. Assuming a simple exponential kinetics the transient charge may be written as,

$$Q(t_1) = Q_{total}\left[1 - \exp\left(\frac{t_1}{\tau}\right)\right], \tag{6.19}$$

where τ and Q_{total} are the time constant and the total charge, respectively. In Q-DLTS, the linear combination of the three transient charges $\Delta Q = Q(t_1) - 3/2Q(2t_1) + 1/2Q(4t_1)$ gives the correlated charge ΔQ, which has a maximum value when the time constant τ is equal to t_1. Then, the amplitude $\Delta Q_{max}(t_1)$ amounts to $0.17 \times Q_{total}$ and is independent of t_1, unless the response is non-exponential.

The temperature dependence of the time constant may be written in the form

$$\tau(T) = \tau_0 \exp\left(\frac{\Delta E}{k_B T}\right), \tag{6.20}$$

with ΔE being the thermal activation energy of the relaxation. The activation energy may be determined an Arrhenius' plot of $\ln(\tau_{max})$ vs $(kT_{max})^{-1}$.

In FCM the first channel at t_0 precedes the trailing edge of the voltage pulse ΔU to set a baseline. In the absence of any steady-state current the first sampled value $Q_C(t_0)$ is zero. The two remaining channels are activated after the trailing edge of the voltage pulse at t_2 and $3t_2$, respectively. The transient charges measured at t_2

and $3t_2$ are due to two processes: a charge $Q_C(0) = C(0)$ due to instantaneous charging of the (geometrical) capacitance $C(0)$, and an excess charge $Q(t)$ due to polarization effects. Since any capacitor will be completely charged for $t > 5\tau$, $Q(t)$ will approach the value Q_{total} at $t_2 > 5\ \tau\ (T)$ and the excess capacitance $\Delta C = Q_{total} / \Delta U$ is added to $C(0)$. On the contrary, if setting $t_2 \ll \tau$, only the instantaneous capacitance $C(0)$ is obtained.

Limiting the discussion to dielectric (dipole) polarization $P(t)$ and assuming simple exponential kinetics the polarization may be written,

$$P(t,T) = P_0(T)\left[1 - \exp\left(-\frac{t}{\tau_D(T)}\right)\right], \tag{6.21}$$

where P_0 is the steadystate polarization. Then the formula for the correlated charge will read

$$\Delta Q = \Delta P(T) = \Delta P(t_1, T) - 3/2\Delta P(2t_1, T) + \tfrac{1}{2}\Delta P(4t_1, T). \tag{6.22}$$

The polarization $P(t,T)$ is related to the permittivity $\Delta\varepsilon = \varepsilon(0) - \varepsilon(\infty)$:

$$P_0(T) = \varepsilon_0(\varepsilon_s - \varepsilon_\infty)\Delta U/d, \tag{6.23}$$

where ε_s and ε_∞ are the static and the high-frequency relative permittivities, respectively. Assuming that ΔU drops completely across a layer with thickness d, the excess capacitance ΔC is given by [288]:

$$\Delta C = \frac{P_0}{\Delta U} = \varepsilon_0(\varepsilon_s - \varepsilon_\infty)/d. \tag{6.24}$$

I. Thurzo and co-workers have applied both techniques, that is Q-DLTS and FCM to investigate Ag/Alq_3/PTCDA/GaAs heterostructures [287]. The results from *in situ* and *ex situ* measurements are presented in Sectio 6.7.1 and 6.7.2, respectively.

6.7.1 *In-situ* diagnostics of Ag/Alq_3/PTCDA/GaAs devices

Figure 6.16 shows a set of C–V curves of a Ag/Alq_3/PTCDA/GaAs taken just after preparation of the structure in vacuum. The data has been taken for different t_2 and pulse duration, while keeping the ratio t_2/pulse duration constant. Changing t_2 form 0.015 ms to 1.5 ms increases the capacitance by almost a factor of 2. No change in the capacitance was found for t_2 longer than 1.5 ms. Therefore, ΔC is determined form the difference in capacitance at smallest t_2 and t_2 = 1.5 ms. The difference is taken at U_g = 0V to reduce the voltage drop across the depletion region of GaAs. Using Eq. (6.24), the lower limit of the permittivity dispersion is $\varepsilon_s - \varepsilon_\infty \geq 2.4$. The observed saturation of ΔC at t_2 = 1.5 ms and pulse = 55 ms is an indication of having a time constant τ_D of the polarization below 1 ms. Compared to inorganic semiconductors the polarization is unusually large.

The *in situ* Q-DLTS characteristics of the Ag/Alq_3/PTCDA/GaAs heterostructure are shown in Figure 6.17. A dominant peak is observed for $t_1 \approx 0.8$ ms, which reaches about 30 $\mu C\,m^{-2}$ and remains practically constant when changing the bias

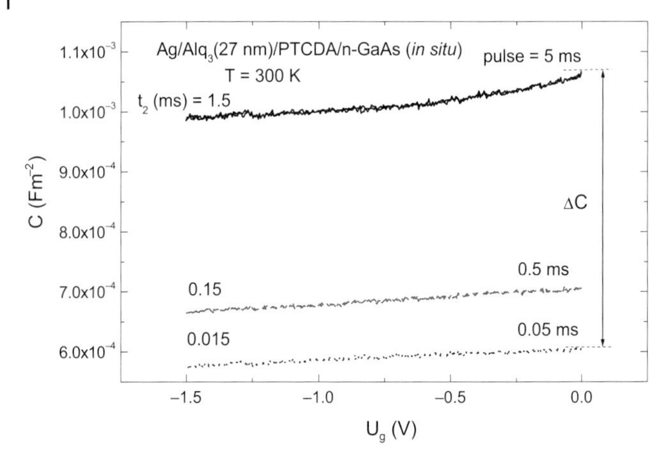

Figure 6.16 *C–V* curves of Ag/Alq$_3$/PTCDA/GaAs heterostructure taken with different t_2. The excess capacitance ΔC is defined as the difference between total capacitance taken at $t_2 \approx 1.5$ ms and the instantaneous capacitance $C(0)$ taken with $t_2 \approx 0.015$ ms [from Reference 287].

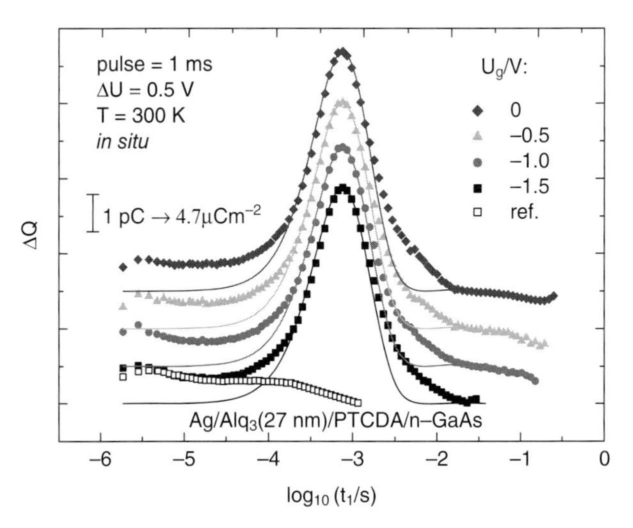

Figure 6.17 *In situ* isothermal Q-DLTS measurements of a Ag/Alq$_3$/PTCDA/GaAs heterostructure (solid symbols). Solid lines correspond to simulated spectra of a polarization characterized by a discrete time constant $\tau = t_{1\,max}$. Open squares show data of a "reference" Ag/PTCDA(28 nm)/n-GaAs sample for comparison [from Reference 287].

U_g. As mentioned above, polarization effects are independent of the bias voltage and the dominant peak is attributed to a polarization in the Alq$_3$ layer. This is corroborated by the data shown for a Ag/PTCDA/GaAs reference sample, where such a peak is completely absent. The position of $t_1 \approx 0.8$ ms confirms a time constant of about 1 ms estimated from FCM. The solid lines correspond to simulated spectra of a polarization characterized by a discrete time constant $\tau = t_{1\,\max}$. The additional contributions to the signal below and above $t_1 \approx 1$ ms likely stem from fast bulk and slow surface states of GaAs, respectively.

6.7.2 *Ex situ* diagnostics of Ag/Alq$_3$/PTCDA/n-GaAs devices

After preparation under vacuum conditions organic thin films have to be transferred out of the vacuum systems for further processing. During this transfer organic thin films are exposed to atmospheric conditions which may change their properties. Here, the type and composition of gas the organic films are exposed to, i.e. air, nitrogen, or noble gases, may play an important role. To investigate the impact of exposure to air on the properties of organic films, the Ag/Alq$_3$/PTCDA/n-GaAs discussed in Section 6.7.1 have been exposed to air and investigated with feedback charge C-V and Q-DLTS using the same measurements settings.

Figure 6.18 shows *C–V* curves taken from a Ag/Alq$_3$/PTCDA/GaAs heterostructure in air. Compared to the *in situ* data shown in Figure 6.16 the geometrical capacitance $C(0)$ is comparable, while the excess capacitance ΔC is by a factor of 2 smaller. In addition, the charging/polarization happens on a smaller time scale, that is, maximal capacitance is already found for 0.15 ms and to determine the geometrical capacitance t_2 has to be as low as 0.006 ms.

This findings are supported by isothermal *Q*-DLTS spectra of the same sample displayed in Figure 6.19. The dominant peak independent of the bias is still observed, but for much shorter time $\tau_D = t_{1\,\max} \approx 5\,\mu$s and smaller amplitude than for the sample measured *in situ*. Since the polarization induced peak shifts to

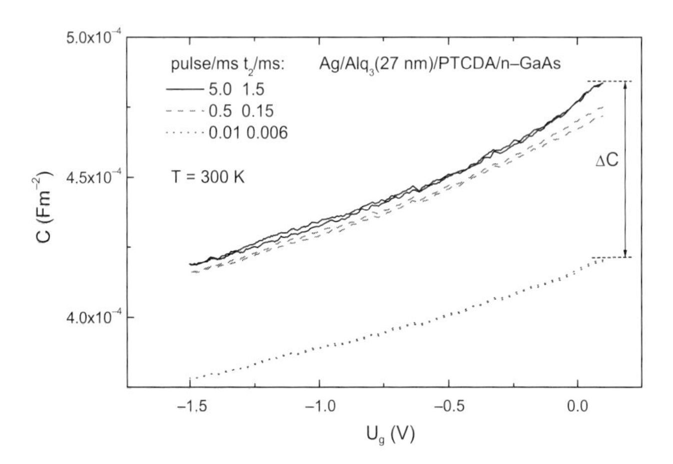

Figure 6.18 C-V curves of Ag/Alq$_3$/PTCDA/GaAs heterostructure exposed to air [from Reference 287].

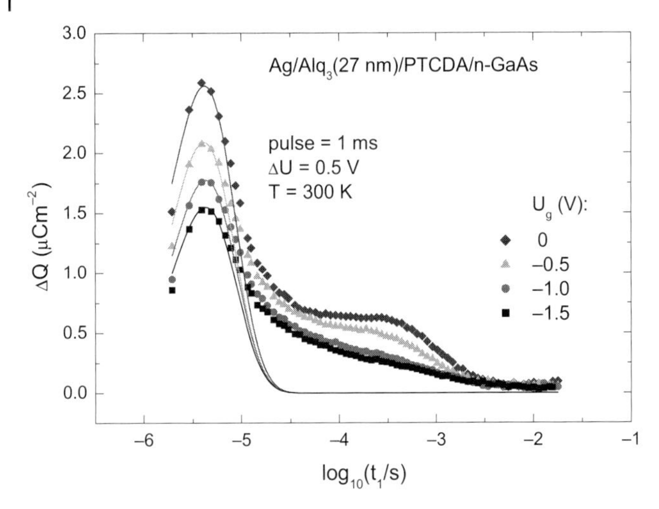

Figure 6.19 *Ex situ* isothermal *Q*-DLTS measurements on a Ag/Alq_3/PTCDA/GaAs heterostructure taken in air. Solid lines correspond to simulated spectra of a polarization characterized by a discrete time constant $\tau = t_{1\,max}$. [from Reference 287].

smaller t_1, another less dominant feature is now observed at $t_{1max} \approx 0.3$ ms. The amplitude of this peak depends on the bias voltage U_g and the peak is therefore attributed to deep levels. It was also observed in the Ag/PTCDA/GaAs referenece sample and is identified as a defect level at the PTCDA/GaAs interface.

The *ex situ* Ag/Alq_3/PTCDA/GaAs is further investigated by recording thermal-scan *Q*-DLTS spectra. Figure 6.20 shows ΔQ as a function of the temperature, recorded for different bias and pulse voltages. In general, three distinct features can be recognized: a broad peak (1) centred around 175 K and a smaller satellite peak (2) at 270 K which both appear for negative and positive pulses, and a sharp feature at about 200 K which is only observed for positive pulses.

The origin of the sharp feature becomes more clear by comparing the two spectra taken with U_g /ΔU settings of −1.5 V/1.5 V and 0 V/−1.5 V. Both settings cover the same voltage region, that is, −1.5 V–0 V, but the pulse direction is reversed. The two pulses excite the same volume element in GaAs, but the sharp feature is only observed for positive pulses. Since a positive polarity of ΔU corresponds to electron capture, the sharp feature is attributed to the 0.3 eV electron trap in the GaAs bulk. For bias pulses of a few volts the electron capture rate is much faster than the electron emission rate [289]. This is the reason why there is no detectable response for $\Delta U = -1.5$ V, the electron emission time constant lying outside the rate window set by t_1^{-1}.

The set of four spectra taken with constant pulse height of 0.5 V but different reverse bias between −1.5 V and 0 V, gives evidence that the residual polarization is enlarged after reducing the potential drop across the space charge region in GaAs, setting the bias U_g to zero. In general, polaruzation effects should be independent of the bias. A possible explanation is, that a redistribution of the applied

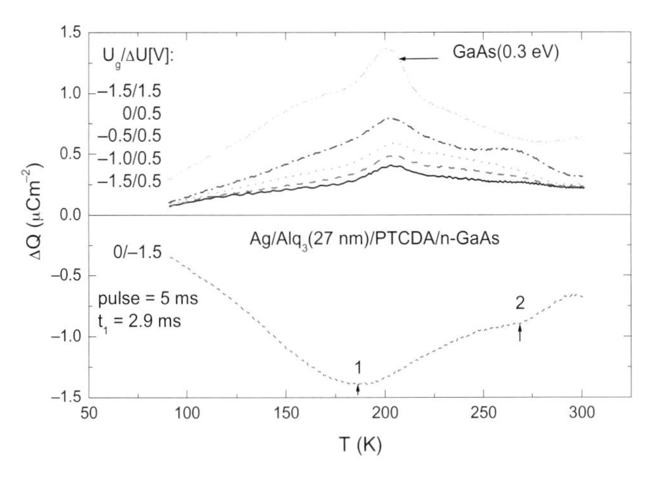

Figure 6.20 The set of thermal *Q*-DLTS scans taken from the $Ag/Alq_3/PTCDA/GaAs$ heterostructure taken in air. Peak 1 and 2 are attributed to polarization while the sharp feature at about 200 K is only observed for positive pulses and is attributed to the 0.3 eV bulk level in *n*-GaAs [from Reference 289].

potential among the individual layers occurs on a time scale of the transient charge processing. Therefore, the broad peak 1 ($T_{max} \approx 185$ K) and the relatively narrow satellite (peak 2) at 270 K can be assigned to Alq_3 and PTCDA, respectively, rather than to traps in the *n*-GaAs substrate.

6.8 Organic-Modified Schottky Diodes for Frequency Mixer Applications

The results presented in the previous sections have implications for the design and manufacturing of device structures used for investigations of high-frequency characteristics. For example, the thickness of the organic film should be of the order of 20 nm and exposure of the organic materials to air should be avoided.

The group of W. Kowalsky at the Technische Universität Braunschweig has investigated the frequency characteristics of Au/PTCDA/GaAs heterostructure diodes [287, 288]. The structure of these diodes is shown in Figure 6.21. The lateral geometry permits one to contact the devices with a ground-signal-ground configured microwave probe. The substrate consisting out of n-type doped GaAs(100) with a doping concentration of $3 \times 10^{18}\,cm^{-3}$ was cleaned by etching in HCl and $1HBr:1CH_3COOH:1K_2Cr_2O_7$(sat.)$:9H_2O$ for 10 min and 8 s, respectively. For the structuring of the ohmic contacts, conventional lift-off technique was used. The ohmic contact was achieved by a AuGe(88 : 12)/Ni/Au (30 nm/20 nm/180 nm) metallization followed by rapid thermal annealing at 300–400 °C in a nitrogen atmosphere. For the structuring of the Schottky contacts, the usually used 500 nm SiO_2

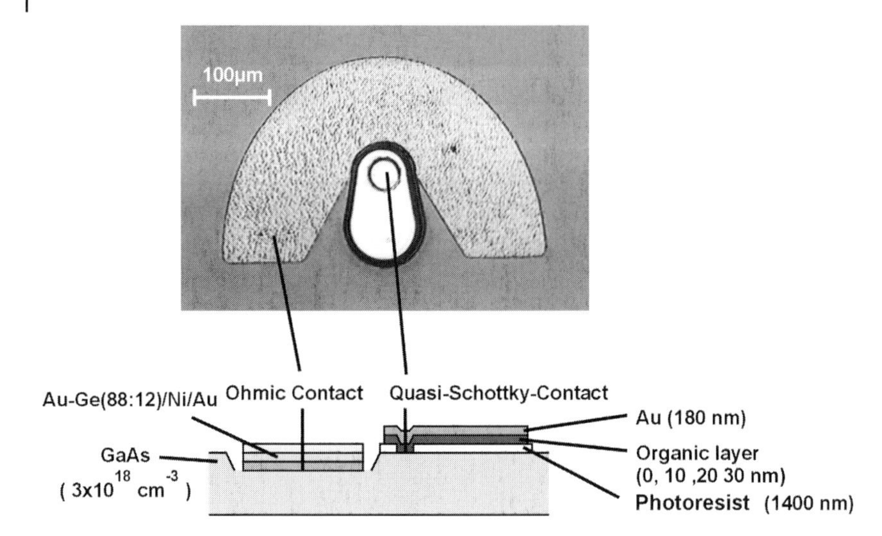

Figure 6.21 Device structure of an Au/PTCDA/GaAs mixer diode. (Courtesy of G. Ginev and W. Kowalsky.)

layer was replaced by a hard-backed photoresist. This has been found to simplify the technology sequence and improve device performance by reducing parasitic device capacitance. In this manner, small active areas with a diameter of 10, 20, 40, and 80 μm were made, which have low junction capacitance and sufficient contact space for testing. Prior to the evaporation of the PTCDA and the top contact metal, the GaAs was cleaned by etching for 20 s in HF (1%). Etching of GaAs in HF has been found to reduce the oxygen contamination on the surface. On the other hand, the surface becomes As rich [289]. Finally, the organic semiconductor and the top metal layer were deposited in a chamber with a base pressure of 5×10^{-6} Pa. In this structure, the organic material is encapsulated in the volume defined by the GaAs substrate, the photoresist, and the Au top layer of 180 nm thickness. This prevents any exposure of the organic film to air.

And indeed, the *IV* characteristics of such a device with PTCDA films of 10, 20, and 30 nm thickness show higher current densities for applied voltages below 0.5 V, indicating a lowering of the effective barrier height due the additional organic layer. Another important result for the high-frequency application is that the junction capacitance is reduced. The optimum dimensions for organic layer thickness and diode diameter are 20 and 40 μm, respectively. Here, the minimum in effective barrier height and an optimum concerning leakage current and junction capacitance is found.

Schottky diodes are used to convert signals from high to low frequencies. The mixing of the two signals, one being a low-frequency signal with frequency f_{LF} and the other a radio frequency signal f_{RF}, takes place in the diode junction. The output is limited by the current flow and the shape of the *IV* characteristics. The ideal device should work in a range, where the output voltage linearly depends on the

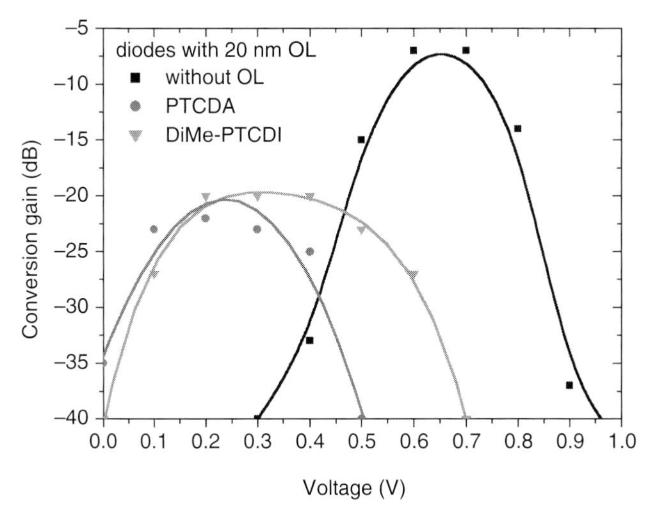

Figure 6.22 Conversion gain of an Ag/PTCDA/GaAs, a Ag/DiMe-PTCDI/GaAs, and a Ag/GaAs Schottky diode as a function of bias. (Courtesy of G. Ginev and W. Kowalsky.)

input voltage of the low-frequency signal. This is achieved by setting an appropriate bias. The frequency of the output signal is given by the difference $f_{LF} - f_{RF}$. The mixing efficiency is measured by the conversion gain, which is defined as the ratio of output power with respect to the power of the radio frequency signal. Figure 6.22 shows a plot of conversion gain as a function of the applied bias for two mixer diodes having a 20-nm PTCDA or DiMe–PTCDI layer, respectively, in comparison with a reference Ag/GaAs diode. The frequencies used for this measurements are $f_{LF} = 500\,\text{MHz}$ and $f_{RF} = 510\,\text{MHz}$, resulting in an output signal with a frequency of 10 MHz. For the reference diode, a maximum conversion gain is found at a bias of 0.65 eV. The organic modification of the mixer diodes reduces the maximum conversion gain, but also shifts it to lower biases. In the case of a 20-nm PTCDA film, acceptable mixing levels are achieved even at 0 V. The frequency limit for the organic-modified Schottky diode has been estimated to be about 40 GHz.

References

1 Burroghs, J.H., Bradley, D.D.C., Brown, A.R., Marks, R.N., Mackay, K., Friend, R.H., Burn, P.L., and Holmes, A.B. (1990) *Nature*, **347**, 539.

2 Tung, C.W., and van Slyke, S.A. (1987) *Appl. Phys. Lett.*, **51**, 913.

3 Frolov, S.V., Liess, M., Lane, P.A., Gellermann, W., Vardeny, Z.V., Ozaki, M., and Yoshino, K. (1997) *Phys. Rev. Lett.*, **78**, 4285.

4 Gadjourova, Z., Andreev, Y.G., Tunstall, D.P., and Bruce, P.G. (2001) *Nature*, **412**, 520.

5 Gelink, G.H., Geuns, T.C.T., and de Leeuw, D.M. (2000) *Appl. Phys. Lett.*, **77**, 1487.

6 Sirringhaus, H., Tessler, N., and Friend, R.H. (1998) *Science*, **280**, 1741.

7 De Leeuw, D.M., Blom, P.W.M., Hart, C.M., Mutsaers, C.M.J., Drury, C.J., Matters, M., and Termeer, H. (1997) *IEDM Techn. Dig.*, 331.

8 Lundquist, P.M., Poga, C., DeVoe, R.G., Jia, Y., Moerner, W.E., Bernal, M.-P., Coufal, H., Grygier, R.K., Hoffnagle, J.A., Jefferson, C.M., MacFairlane, R.M., Shelby, R.M., and Sincerbox, G.T. (1996) *Opt. Lett.*, **21**, 890.

9 Pei, Q., and Inganäs, O. (1993) *Synth. Mat.*, **57**, 3730.

10 Sariciftci, N.S., Braun, D., Zhang, C., Srdanov, V.I., Heeger, A.J., Stucky, G., and Wudl, F. (1993) *Appl. Phys. Lett.*, **62**, 585.

11 Sheats, J.R., Antoniadis, H., Hueschen, M., Leonard, M.W., Milller, J., Moon, R., Roitman, D., and Stocking, A. (1996) *Science*, **273**, 884.

12 Gu, G., and Forrest, S.R. (1998) *IEEE J. Sel. Top. Quantum Electron.*, **4**, 83.

13 Forrest, S.R. (1997) *Chem. Rev.*, **97**, 1793.

14 Burrows, P.E., and Forrest, S.R. (1993) *Appl. Phys. Lett.*, **62**, 3102.

15 Taylor, R.T., Burrows, P.E., and Forrest, S.R. (1997) *IEEE Photonics Technol. Lett.*, **9**, 365.

16 Dunn, A.W., Moriaty, P., Upward, M.D., and Beton, P.H. (1996) *Appl. Phys. Lett.*, **69**, 506.

17 Maas, S.A. (1986) *Microwave Mixers*, Artech House, Norwood, MA.

18 Karl, N., and Günther, C.H. (1999) *Cryst. Res. Technol.*, **34**, 243.

19 Sandoz (1957) DE-AS 1 071 280.

20 Hoechst (1980) DE-OS 3 008 420.

21 Tröster, H. (1983) *Dyes Pigm.*, **4**, 171.

22 Pochettino, A. (1906) *Atti Accad. Naz. Lincei, Cl. Sci. Fis. Mat. Nat. Rend.*, **15**, 355.

23 Koenigsberger, J., and Schilling, K. (1910) *Ann. Phys.*, **32**, 179.

24 Volmer, M. (1913) *Ann. Phys.*, **40**, 775.

25 Lüth, H. (1993) *Surfaces and Interfaces of Solids*, 2nd edn, vol. **15**, Springer Series in Surface Science, Springer, Berlin.

26 Mönch, W. (1995) *Semiconductor Surfaces and Interfaces*, 3rd edn, vol. **26**, Springer Series in Surface Science, Springer, Berlin.

27 O. Madelung (editor), (1982) *Landolt-Börnstein: Numerical Data and Functional Relationships in Science and Technology*, vol. III/17, Springer, Berlin.

Low Molecular Weight Organic Semiconductors. Thorsten U. Kampen

ISBN: 978-3-527-40653-1

28 Heine, V. (1965) *Phys. Rev.*, **138**, A1689.
29 Maue, A.W. (1935) *Z. Phys.*, **94**, 717.
30 Kampen, T.U., Koenders, L., Smit, K., Rückschloss, M., and Mönch, W. (1991) *Surf. Sci.*, **242**, 314.
31 Stiles, K., Kahn, A., Kilday, D., and Magaritondo, G. (1987) *J. Vac. Sci. Technol. B*, **5**, 987.
32 Cao, R., Miyano, K., Kendelewicz, T., Chin, K.K., Lindau, I., and Spicer, W.E. (1987) *J. Vac. Sci. Technol. B*, **5**, 998.
33 Mönch, W. (1988) *J. Vac. Sci. Technol. B*, **6**, 1270.
34 Pauling, L.N. (1960) *The Nature of the Chemical Bond*, Cornell University, Ithaca, NY.
35 Hanney, N.B., and Smith, C.P. (1946) *J. Am. Chem. Soc.*, **68**, 171.
36 Topping, J. (1927) *Proc. Roy. Soc. A*, **114**, 67.
37 Grundner, M., and Jacobs, H. (1986) *Appl. Phys. A*, **39**, 73.
38 Chabal, Y.J. (1989) *J. Vac. Sci. Technol. A*, **7**, 2104.
39 Stockhausen, A., Kampen, T.U., and Mönch, W. (1992) *Appl. Surf. Sci.*, **56–58**, 795.
40 Flores, F., and Miranda, R. (1994) *Adv. Mat.*, **6**, 540.
41 Saiz-Pardo, R., Pérez, R., García-Vidal, F.J., Whittle, R., and Flores, F. (1999) *Surf. Sci.*, **426**, 26.
42 Kampen, T.U., Schmitsdorf, R., and Mönch, W. (1995) *Appl. Phys. A*, **60**, 391.
43 Hohenecker, St., Kampen, T.U., Zahn, D.R.T., and Braun, W. (1998) *J. Vac. Sci. Technol. B*, **16**, 2317.
44 Hohenecker, St, Kampen, T.U., Werninghaus, T., Zahn, D.R.T., and Braun, W. (1999) *Appl. Surf. Sci.*, **142**, 28.
45 Mönch, W., Schmitsdorf, R., Kampen, T.U., and Stockhausen, A. (1995) *Proceedings of 22nd International Conference on the Physics of Semiconductors* (ed. J. Lockwood) World Scientific, Singapore, p. 576.
46 Zahn, D.R.T., Kampen, T.U., Hohenecker, S., and Braun, W. (2000) *Vacuum*, **57**, 139.
47 Kampen, T.U., Zahn, D.R.T., Braun, W., Gonzáles, C., Benito, I., Ortega, J., Jurczyszyn, L., Blanco, J.M., Pérez, R., and Flores, F. (2003) *Appl. Surf. Sci.*, **212–213**, 850–855.
48 Sandroff, C.J., Nottenburg, R.N., Bischoff, T.C., and Bhat, R. (1987) *Appl. Phys. Lett.*, **51**, 33.
49 Kamiyama, S., Mori, Y., Takahashi, Y., and Ohnaka, K. (1987) *Appl. Phys. Lett.*, **58**, 2595.
50 Howard, A.J., Ashby, C.I.H., Lott, J.A., Schneider, R.P., and Coreless, R.F. (1996) *J. Vac. Sci. Technol. A*, **12**, 1063.
51 Bessolov, V.N., Lebedev, M.V., Shernyakov, Y.M., and Tsarenkov, B.V. (1997) *Mater. Sci. Eng. B*, **44**, 380.
52 Anderson, G.W., Hanf, M.C., Qin, X.R., Norton, P.R., Myrtle, K., and Heinrich, B. (1996) *Surf. Sci.*, **346**, 145.
53 Hirose, Y., Forrest, S.R., and Kahn, A. (1995) *Phys. Rev. B*, **52**, 14040.
54 Kampen, T.U., Salvan, G., Tenne, D., and Zahn, D.R.T. (2001) *Appl. Surf. Sci.*, **175–176**, 326.
55 Pashley, M.D., and Li, D. (1994) *J. Vac. Sci. Technol. A*, **12**, 1848.
56 Takatani, S., Kikawa, T., and Nakazawa, M. (1992) *Phys. Rev. B*, **59**, 8498.
57 Pashley, M.D. (1989) *Phys. Rev. B*, **40**, 10481.
58 Gundel, S., and Faschinger, W. (1999) *Phys. Rev. B*, **59**, 5602.
59 Gonzalez, C., Benito, I., Ortega, J., Jurczyszyn, L., Blanco, J.M., Perez, R., Flores, F., Kampen, T.U., Zahn, D.R.T., and Braun, W. (2004) *J. Phys.: Condens. Matter*, **16**, 2187.
60 Li, Z.S., Cai, W.Z., Su, R.Z., Dong, G.S., Huang, D.M., Ding, X.M., Hou, X.Y., and Wang, X. (1994) *Appl. Phys. Lett.*, **64** (25), 3425.
61 LeLay, G., Mao, D., Kahn, A., Hwu, Y., and Margaritondo, G. (1991) *Phys. Rev. B*, **43**, 14301.
62 Vitomirov, I.M., Raisanen, A.D., Finnefrock, A.C., Viturro, R.E., Brillson, L.J., Kirchner, P.D., Pettit, G.D., and Woodall, J.M. (1992) *J. Vac. Sci. Technol. B*, **10**, 1898.
63 Hirsch, G., Krüger, P., and Pollmann, J. (1998) *Surf. Sci.*, **402–404**, 778.
64 Demkov, A.A., Ortega, J., Sankey, O.F., and Grumbach, M.P. (1995) *Phys. Rev. B*, **52**, 1618.
65 Braun, F. (1874) *Pogg. Ann. Phys.*, **153**, 556.

66 Schottky, W. (1938) *Naturwissenschaften*, **26**, 843.

67 Schottky, W. (1940) *Physik. Zeitschr.*, **41**, 570.

68 Mott, N.F. (1938) *Proc. Camb. Philol. Soc.*, **34**, 568.

69 Anderson, R.L. (1962) *Solid State Electron.*, **5**, 341.

70 Schweikert, H. (1939) *Verhandl. Phys. Ges.*, **3**, 99.

71 Bardeen, J. (1947) *Phys. Rev.*, **71**, 717.

72 Mönch, W. (1987) *Phys. Rev. Lett.*, **58**, 1260.

73 Mönch, W. (1996) *J. Vac. Sci. Technol. B*, **14**, 2985.

74 Mönch, W. (1989) *Gallium Arsenide Technology*, vol. II (ed. D.K. Ferry), Sams, Carmel, IN, p. 139.

75 Mönch, W. (1986) *Festkörperprobleme*, vol. 26 (ed. P. Grosse), (Adv. in Solid State Phys.), Vieweg, Braunschweig, p. 67.

76 Mönch, W. (1996) *Appl. Surf. Sci.*, **92**, 367.

77 de Boer, F.R., Boom, R., Mattens, W.C.M., Miedema, A.R., and Nielsen, A.K. (1988) *Cohesion in Metals* (eds F.R. de Boer and D.G. Pettefos), North-Holland, Eindhoven, p. 699.

78 Tersoff, J. (1984) *Phys. Rev. Lett.*, **52**, 465.

79 Tersoff, J. (1986) *J. Vac. Sci. Technol. B*, **4**, 1066.

80 Mönch, W. (1996) *J. Appl. Phys.*, **80**, 5076.

81 Baldereschi, A. (1973) *Phys. Rev. B*, **7**, 5212.

82 Penn, D.R. (1962) *Phys. Rev.*, **128**, 2093.

83 Rohlfing, M., Krüger, P., and Pollmann, J. (1993) *Phys. Rev. B*, **48**, 17791.

84 Rohlfing, M., Krüger, P., and Pollmann, J. (1995) *Phys. Rev. Lett.*, **75**, 3489.

85 Tung, R.T. (1992) *Phys. Rev. B*, **45**, 13509.

86 Sullivan, J.P., Tung, R.T., Pinto, M.R., and Graham, W.R. (1991) *J. Appl. Phys.*, **70**, 7403.

87 Rau, U., Güttler, H.H., and Weerner, J.H. (1992) *Mater. Res. Soc. Symp. Proc.*, **260**, 245.

88 Schmitsdorf, R.F., Kampen, T.U., and Mönch, W. (1995) *Surf. Sci.*, **324**, 249.

89 Esser, N., Reckzügel, M., Srama, R., Resch, U., Zahn, D.R.T., Richter, W., Stephens, C., and Hünnermann, M. (1990) *J. Vac. Sci. Technol. B*, **8**, 680.

90 Cao, R., Miyamo, K., Kendelewicz, T., Lindau, I., and Spicer, W.E. (1988) *Surf. Sci.*, **206**, 413.

91 Hardikar, S., Hudait, M.K., Modak, P., Krupanidhi, S.B., and Padha, N. (1999) *Appl. Phys. A*, **68**, 49.

92 Bhuiyan, A.S., Martinez, A., and Esteve, D. (1988) *Thin Solid Films*, **161**, 93.

93 DiDio, M., Cola, A., Lupo, M.G., and Vasanelli, L. (1995) *Solid State Electron.*, **38**, 1923.

94 Arulkumaran, S., Arokiaraj, J., Dharmarasu, N., and Kumar, J. (1996) *Nucl. Instrum. Methods. Phys. Res. B*, **116**, 519.

95 Nuhoglu, Ç., Ayyíldíz, E., Saglam, M., and Türüt, A. (1998) *Appl. Surf. Sci.*, **135**, 350.

96 Hackham R., and Harrop P. (eds), (1972) *IEEE Trans.* Electron Devices **ED-19**, 1231.

97 Nathan, M., Soshani, Z., Ashkinazi, G., Meyler, B., and Zolotareveski, O. (1996) *Solid State Electron.*, **39**, 1457.

98 Dharmarusu, N., Arulkumaran, S., Sumathi, R.R., Jayavel, P., Kumar, J., Magudapathy, P., and Nair, K.G.M. (1998) *Nucl. Instrum. Methods. Phys. Res. B*, **140**, 119.

99 Barnars, W.O., Myburg, G., Auret, F.D., Goodman, S.A., and Meyer, W.E. (1996) *J. Electron. Mater.*, **25**, 1695.

100 Hübers, H.-W., and Röser, H.P. (1998) *J. Appl. Phys.*, **84**, 5326.

101 Waldrop, J.R. (1984) *J. Vac. Sci. Technol. B*, **2**, 445.

102 Spicer, W.E., Gregory, P.E., Chye, P.W., Babaola, J.A., and Sukegawa, T. (1975) *Appl. Phys. Lett.*, **27**, 617.

103 Grunwald, F. (1987) Diploma thesis, Universität Duisburg (unpublished).

104 Mao, D., Santos, M., Shayegan, M., Kahn, A., Lay, G.Le, Hwu, Y., Margaritondo, G., Florez, L.T., and Harbison, J.P. (1992) *Phys. Rev. B*, **45**, 1273.

105 Prietsch, M., Domke, M., Laubschat, C., Mandel, T., Xue, C., and Kaindl, G. (1989) *Z. Phys. B – Condensed Matter*, **74**, 21.

106 Hohenecker, St., Patchett, A., Drews, D., and Zahn, D.R.T. (1996) *BESSY Jahresbericht*, 323.

107 Hohenecker, St. (2001) Chalcogen modification of GaAs(100) surfaces and metal/GaAs(100) contacts, PhD thesis, Chemnitz University of Technology, Germany.

108 Hohenecker, St., Drews, D., Lübbe, M., Zahn, D.R.T., and Braun, W. (1998) *Appl. Surf. Sci.*, **123**, 585.

109 Hohenecker, St., Drews, D., Werninghaus, T., Zahn, D.R.T., and Braun, W. (1998) *J. Electron Spectrosc. Relat. Phenom.*, **96**, 97.

110 Hohenecker, St., Kampen, T.U., Braun, W., and Zahn, D.R.T. (1999) *Surf. Sci.*, **433–435**, 347.

111 Iqbal, Z., Ivory, D.M., and Eckhardt, H. (1988) *Mol. Cryst. Liq. Cryst.*, **158B**, 337.

112 Wüsten, J., Ertl, T.H., Lach, S., and Ziegler, C.H. (2005) *Appl. Surf. Sci.*, **252**, 104.

113 Maliakal, A., Raghavachari, K., Katz, H., Chandross, E., and Siegrist, T. (2004) *Chem. Mater.*, **16**, 4980–4986.

114 Rolin, C., Steudel, S., Myny, K., Cheyns, D., Verlaak, S., Genoe, J., and Heremans, P. (2006) *Appl. Phys. Lett.*, **89**, 203502.

115 Yang, F., Shtein, M., and Forrest, S.R. (2005) *Nat. Mat.*, **4**, 37.

116 Möbus, M., Karl, N., and Kobayashi, T. (1992) *J. Cryst. Growth*, **116**, 495.

117 Lovinger, A.J., Forrest, S.R., Kaplan, M.L., Schmidt, P.H., and Venkatesan, T. (1984) *J. Appl. Phys.*, **55**, 476.

118 Lovinger, A.J., Forrest, S.R., Kaplan, M.L., Schmidt, P.H., and Venkatesan, T. (1983) *Bull. Am. Phys. Soc.*, **28**, 363.

119 Chizhov, I., Kahn, A., and Scoles, G. (2000) *J. Cryst. Growth*, **208**, 449.

120 Ogawa, T., Kuwamotot, K., Isoda, S., Kobayashi, T., and Karl, N. (1999) *Acta Crystallogr B*, **55**, 123.

121 Hädicke, E., and Graser, F. (1986) *Acta Crystallogr. C*, **42**, 189.

122 Umbach, E., Glöckner, K., and Sokolowski, M. (1998) *Surf. Sci.*, **402–404**, 20.

123 Schmitz-Hübsch, T., Fritz, T., Sellam, F., Straub, R., and Leo, K. (1997) *Phys. Rev. B*, **55**, 7972.

124 Kendrick, C., Kahn, A., and Forrest, S.R. (1996) *Appl. Surf. Sci.*, **104/105**, 586.

125 Seidel, C., Poppensieker, J., and Fuchs, H. (1998) *Surf. Sci.*, **408**, 223.

126 Tautz, F.S., Eremtchenko, M., Schaefer, J.A., Sokolowski, M., Shklover, V., and Umbach, E. (2002) *Phys. Rev. B*, **65**, 125405.

127 Taborski, J., Väterlein, P., Dietz, H., Zimmermann, U., and Umbach, E. (1995) *J. Electron Spectrosc. Relat. Phenom.*, **75**, 129.

128 Zimmermann, U., Schnitzler, G., Karl, N., Dudde, R., and Koch, E.E. (1989) *Thin Solid Films*, **175**, 85.

129 Zimmermann, U., Schnitzler, G., Schneider, M., Kaiser, M., Herde, R., Wüstenhagen, V., Taborski, J., Karl, N., Dudde, R., Koch, E.E., and Umbach, E. (1990) *Vaccum*, **41**, 1795.

130 Kawaguchi, T., Tada, H., and Koma, A. (1994) *J. Appl. Phys.*, **75**, 1486.

131 Kendrick, C., and Kahn, A. (1997) *J. Cryst. Growth*, **181**, 181.

132 Hirose, Y., Chen, W., Haskal, E.I., Forrest, S.R., and Kahn, A. (1994) *Appl. Phys. Lett*, **64**, 3482.

133 Möbus, M. (1992) PhD thesis, University of Stuttgart.

134 Nowakowski, R., Seidel, C., and Fuchs, H. (2001) *Phys. Rev. B*, **63**, 195418.

135 Yanagi, H., Toda, Y., and Noguchi, T. (1995) *Jpn. J. Appl. Phys.*, **34**, 3808.

136 Lifshitz, E., Kaplan, A., Ehrenfreund, E., and Meissner, D. (1999) *Chem. Phys. Lett.*, **300**, 626.

137 Schäfer, A.H., Seidel, C., and Fuchs, H. (1998) *Thin Solid Films*, **379**, 176.

138 Seidel, C., Schäfer, A.H., and Fuchs, H. (2000) *Surf. Sci.*, **459**, 310.

139 Fowler, R.H., and Nordheim, L.W. (1928) *Proc. Roy. Soc. A*, **119**, 173.

140 Nicoara, N., Román, E.R., Gomez-Rodriguez, J.M., Martin-Gago, J.A., and Mendez, J. (2006) *Org. Electron.*, **7**, 287–294.

141 Perdew, J.P., and Zunger, A. (1981) *Phys. Rev. B*, **23**, 5048.

142 Fenter, P., Eisenberger, P., Burrows, P., Forrest, S.R., Ordejon, K.S.P., Artacho, E., and Soler, J.M. (1996) *Phys. Rev. B*, **53**, R10441.

143 Soler, J.M., Artacho, E., Gale, J., Garcý'a, A., Junquera, J., Ordejon, P., and Sanchez-Portal, D. (2002) *J. Phys., Condens. Matter.*, **14**, 2745.

144 Nicoara, N., Custance, O., Granados, D., Garcia, J.M., Gomez-Rodriguez, J.M., Baro, A.M., and Mendez, J. (2003) *J. Phys., Condens. Matter.*, **15**, S2619–S2629.

145 Binnig, G., Quate, C.F., and Gerber, C. (1986) Atomic force microscope. *Phys. Rev. Lett.*, **56**, 930.

146 Moreno-Herrero, F., de Pablo, P.J., Colchero, J., Gómez-Herrero, J., and Baró, A.M. (2000) *Surf. Sci.*, **453**, 152.

147 Bragg, W.L. (1913) *Proc. Camb. Philols. Soc.*, **17**, 43.

148 Akers, K., Aroca, R., Hor, A., and Loutfy, R.O. (1987) *J. Phys. Chem.*, **91**, 2954.

149 Salvan, G., Tenne, D., Das, A., Kampen, T.U., and Zahn, D.R.T. (2000) *J. Org. Electron.*, **1**, 49.

150 Cullity, B.D. (1956) *Elements of X-Ray Diffraction*, Addison-Wesley, Reading, MA.

151 Fenter, P., Schreiber, F., Zhou, L., Eisenberg, P., and Forrest, S.R. (1997) *Phys. Rev. B*, **56**, 3046.

152 Parratt, L.G. (1954) *Phys. Rev.*, **95**, 359.

153 Segmuller, A., Noyan, I.C., and Speriosu, V.S. (1989) *Prog. Cryst. Growth Charact.*, **18**, 21.

154 Möbus, M., and Karl, N. (1992) *Thin Solid Films*, **215**, 213.

155 Möbus, M., and Karl, N. (1992) *Thin Solid Films*, **215**, 213.

156 Kampen, T.U., Parian, A.M., Rossow, U., Park, S., Salvan, G., Wagner, T.H., Friedrich, M., and Zahn, D.R.T. (1999) *Phys. Stat. Sol. A*, **188**, 1307.

157 Ha, J.S., Park, S.-J., Kim, S.-B., and Lee, E.-H. (1995) *J. Vac. Sci. Technol. A*, **13**, 646.

158 Mannsfeld, S., Toerker, M., Schmitz-Hübsch, T., Sellam, F., Fritz, T., and Leo, K. (2001) *Org. Electron.*, **2**, 121.

159 Ya, I., and Frenkel (1931) *Phys. Rev.*, **37**, 1276.

160 Wannier, G.H. (1937) *Phys. Rev.*, **52**, 191.

161 Zang, D.Y., So, F.F., and Forrest, S.R. (1991) *Appl. Phys. Lett.*, **59**, 823.

162 Friedrich, M., Wagner, T.H., Salvan, G., Park, S., Kampen, T.U., and Zahn, D.R.T. (2002) *Appl. Phys. A*, **75**, 501.

163 Klebe, G., Graser, F., Hädicke, E., and Berndt, G. (1989) *Acta Crystallogr. C*, **45**, 69.

164 Hoffmann, M. (2009) *Frenkel and Charge-Transfer Excitons in Quasi-One-Dimensional Molecular Crystals with Strong Intermolecular Orbital Overlap*, PhD thesis, Technische Universität Dresden.

165 Scholz, R., Vragovic, I., Kobitski, A.Y., Salvan, G., Kampen, T.U., Schreiber, M., and Zahn, D.R.T. (2002) Proc. Scuola Internationale di Fizika “Enrico Fermi”, course CXLIX, Organic Nanostructures: Science and Applications, Course CXLIX, edited by VM Agranovich and GC La Rocca IOS, Amsterdam, pp. 379.

166 Frisch, M.J., Trucks, G.W., Schlegel, H.B., Scuseria, G.E., Robb, M.A., Cheeseman, J.R., Zakrzewski, V.G., Montgomery, J.A., Stratmann, R.E., Burant, J.C., Dapprich, S., Millam, J.M., Daniels, A.D., Kudin, K.N., Strain, M.C., Farkas, O., Tomasi, J., Barone, V., Cossi, M., Cammi, R., Mennucci, B., Pomelli, C., Adamo, C., Clifford, S., Ochterski, J., Petersson, G.A., Ayala, P.Y., Cui, Q., Morokuma, K., Malick, D.K., Rabuck, A.D., Raghavachari, K., Foresman, J.B., Cioslowski, J., Ortiz, J.V., Stefanov, B.B., Liu, G., Liashenko, A., Piskorz, P., Komaromi, I., Gomperts, R., Martin, R.L., Fox, D.J., Keith, T., Al-Laham, M.A., Peng, C.Y., Nanayakkara, A., Gonzalez, C., Challacombe, M., Gill, P.M.W., Johnson, B.G., Chen, W., Wong, M.W., Andres, J.L., Head-Gordon, M., Replogle, E.S., and Pople, J.A. (1998) *Gaussian 98* (Revision A.1), Gaussian, Inc., Pittsburgh PA.

167 Kaiser, R., Friedrich, M., Schmitz-Hübsch, T., Sellam, F., Kampen, T.U., Leo, K., and Zahn, D.R.T. (1999) *Fresenius J. Anal. Chem.*, **363**, 189.

168 Bulovic, V., Burrows, P.E., Forrest, S.R., Cronin, J.A., and Thompson, M.E. (1996) *Chem. Phys.*, **210**, 1.

169 Hill, I.G., Kahn, A., Soos, Z.G., and Pascal, R.A., Jr. (2000) *Chem. Phys. Lett.*, **327**, 181.

170 Silinsh, E.A., and Cápek, V. (1994) *Organic Molecular Craystals: Interaction, Loalization, and Transport Phenomena*, AIP Press, New York.

171 Slattery, D.K., Linkous, C.A., and Gruhn, N. (2000) *Polym. Prepr.*, **41**, 866.
172 Salaneck, W.R. (1978) *Phys. Rev. Lett.*, **40**, 60.
173 Tsiper, E.V., Soos, Z.G., Gao, W., and Kahn, K. (2002) *Chem. Phys. Lett.*, **360**, 47.
174 Kobitski, A.Yu, Scholz, R., Vragovic, I., Wagner, H.P., Zahn, D.R.T. (2002) *Phys. Rev. B*, **66**, 153204.
175 Kobitski, A.Yu, Scholz, R., Wagner, H.P., Zahn, D.R.T. (2003) *Phys. Rev. B* **68**, 155201.
176 Kobitski, A.Yu (2003) Ph.D. Thesis, Chemnitz. Time-resolved Photoluminescence and Theoretical Study of Excitons in PTCDA.
177 Vragovic, I., Scholz, R., Schreiber, M. (2002) *Europhys. Lett.*, **57**, 288.
178 Scholz, R., Vragovic, I., Kobitski, A.Y., Schreiber, M., Wagner, H.P., Zahn, D.R.T., (2002) *Phys. Stat. Sol. B*, **234**, 402.
179 Scholz, R., Vragovic, I., Kobitski, A.Y., Salvan, G., Kampen, T.U., Schreiber, M., Zahn, D.R.T. (2002) in: V.M. Agranovich, G.C. La Rocca (Ed.), *Proceedings of Course CXLIX: Organic Semiconductors: Science and Application, International School of Physics "E. Fermi", IOS Press, Amsterdam*, pp. 379–403.
180 Kobitski, A.Yu, Scholz, R., Salvan, G., Kampen, T.U., Wagner, H.-P., and Zahn, D.R.T. (2003) *Appl. Sur. Sci.*, **212–213**, 428–432.
181 Loudon, R. (1983) *The Quantum Theory of Light*, Clarendon Press, Oxford, (Chapter 2.4).
182 Turro, N.J. (1991) *Modern Molecular Photochemistry*, University Science Book, Mill Valley, CA, (Chapter 5.7).
183 Scholz, R., Kobitski, A.Yu, Kampen, T.U., Schreiber, M., Zahn, D.R.T., Jungnickel, G., Sternbrg, M., Frauenheim, T. (2000) *Phys. Rev. B*, **61**, 13659.
184 Salvan, G., Tenne, D.A., Das, A., Kampen, T.U., Zahn, D.R.T. (2000) *Org. Electron.*, **1**, 49.
185 Johs, B., Woollam, J. A., Herzinger, C., Hilfiker, J., Synowicki, R., and Bungay, C. (1999) *SPIE* Proceedings, *CR72*, 29–58.
186 Johs, B., and Herzinger, C. (1995) *Guide to Using WVASE32*, J.A. Woolan Co., Inc, Lincoln.
187 Salvan, G. (2003) Metal/Organic/Inorganic Semiconductor Heterostructures Characterized By Vibrational Spectroscopies, PhD thesis, Chemnitz.
188 Scholz, R., Friedrich, M., Salvan, G., Kampen1, T.U., Zahn, D.R.T., and Frauenheim, T. (2003) *J. Phys., Condens. Matter.*, **15**, S2647–S2663.
189 Turrell, G. (1972) *Infrared and Raman Spectra of Crystals*, Chapter 5, Academic Press, London.
190 Azzam, R.M.A., and Bashara, N.M. (1997) *Ellipsometry and Polarized Light*, North-Holland Personal Library.
191 Adachi, S. (1999) *Optical Constants of Crystalline and Amorphous Semiconductors*, Kluwer Academic Publishers, Dordrecht.
192 Faltermeier, D., Gompf, B., Dressel, M., Tripathi, A.K., and Pflaum, J. (2006) *Phys. Rev. B*, **74**, 125416.
193 de Boer, R.W.I., Gershenson, M.E., Morpurgo, A.F., and Podzorov, V. (2004) *Phys. Status Solidi A*, **201**, 1302.
194 Aspnes, D.E., Harbison, J.P., Studna, A.A., Florez, L.T., and Kelly, M.K. (1988) *J. Vac. Sci. Technol. A*, **6**, 1327.
195 Kampen, T.U., Rossow, U., Schumann, M., Park, S., and Zahn, D.R.T. (2000) *J. Vac. Sci. Technol. B*, **18**, 2077.
196 Aspnes, D.E., and Studna, A.A. (1983) *Phys. Rev. B*, **27**, 985.
197 Hughes, G., Springer, C., Resch, U., Esser, N., and Richter, W. (1995) *J. Appl. Phys.*, **78**, 1948.
198 Berkovits, V.L., and Paget, D. (1992) *Appl. Phys. Lett.*, **61**, 1835.
199 Hill, I.G., Rajagopal, A., Kahn, A., and Hu, Y. (1998) *Appl. Phys. Lett.*, **73**, 662.
200 Narioka, S., Ishii, H., Yoshimura, D., Sei, M., Ouchi, Y., Seki, K., Hasegawa, S., Miyazaki, T., Harima, Y., and Yamashita, K. (1995) *Appl. Phys. Lett.*, **67**, 1899.
201 Ishii, H., and Seki, K. (2002) Energy level alignment at oragnic-metal interfaces, in *Conjugated Polymer and Molecular Interfaces* (eds W.R. Salaneck, K. Seki, A. Kahn, and J.-J. Pireaux), Marcel Dekker, New York, pp. 293–249.

202 Chen, Y.C., Cunningham, J.E., and Flynn, C.P. (1984) *Phys. Rev. B*, **30**, 7217.

203 Zangwill, A. (1988) *Physics at Surfaces*, Cambridge University Press, Cambridge, UK, p. 185.

204 Lang, N.D. (1981) *Phys. Rev. Lett.*, **46**, 842.

205 Lang, N.D., and Williams, A.R. (1982) *Phys. Rev. B*, **25**, 2940.

206 Less, K.J., and Wison, E.G. (1973) *J. Phys. C*, **6**, 3110.

207 Fujihira, M., and Inokuchi, H. (1972) *Chem. Phys. Lett.*, **17**, 554.

208 Ito, E., Oji, H., Ishii, H., Oichi, K., Ouchi, Y., and Seki, K. (1998) *Chem. Phys. Lett.*, **287**, 137.

209 Ishii, H., Morikawa, E., Tang, S.J., Yoshimura, D., Ito, E., Okudeira, K.K., Miyamae, T., Hasegawa, S., Sprunger, P.T., Ueno, N., Seki, K., and Saile, V. (1999) *J. Electron Spectrosc. Relat. Phenom.*, **101**, 559.

210 Yoshimura, D., Ishii, H., Ouchi, Y., Ito, E., Miyamae, T., Hasegawa, S., Okudeira, K.K., Ueno, N., and Seki, K. (1999) *Phys. Rev. B*, **60**, 9046.

211 Umbach, E., and Fink, R. (2002) How to control the properties of interfaces and thin films of organic molecules? in *Conjugated Polymer and Molecular Interfaces* (eds W.R. Salaneck, K. Seki, A. Kahn, and J.-J. Pireaux), Marcel Dekker, New York, p. 241.

212 Mui, C., Han, J.H., Wang, G.T., Musgrave, C.B., and Bent, S.F. (2002) *J. Am. Chem. Soc.*, **124**, 4027.

213 Wang, G.T., Mui, C., Tannaci, J.F., Filler, M.A., Musgrave, C.B., and Bent, S.F. (2003) *J. Phys. Chem. B*, **107**, 4982.

214 Hill, I.G., Makinen, A.J., and Kafa, Z.H. (2000) *Appl. Phys. Lett.*, **77**, 1825.

215 Okudeira, K.K., Hasegawa, S., Ishii, H., Seki, K., Harada, Y., and Ueno, N. (1999) *J. Appl. Phys.*, **85**, 6453.

216 Park, Y., Choong, V.-E., Hsieh, B.R., Tang, C.W., Wehrmeister, T., Müllen, K., and Gao, Y. (1997) *J. Vac. Sci. Technol. A*, **15**, 2574.

217 Park, Y., Choong, V.-E., Ettedgui, E., Gao, Y., Hsieh, B.R., Wehrmeister, T., and Müllen, K. (1996) *Appl. Phys. Lett.*, **69**, 1080.

218 Ueno, N., Sugita, K., Koga, O., and Suzuki, S. (1983) *Jpn. J. Appl. Phys.*, **22**, 1613.

219 Rajagopal, A., and Kahn, A. (1998) *J. Appl. Phys. Phys.*, **84**, 355.

220 Johansson, N., Osada, T., Strafström, S., Salanek, W.R., Parente, V., dos Santos, D.A., Crispin, X., and Bredas, L.J. (1999) *J. Chem. Phys.*, **11**, 2157.

221 Choong, V.-E., Mason, M.G., Tang, C.W., and Gao, Y. (1998) *Appl. Phys. Lett.*, **72**, 2689.

222 Hill, I.G., Schwartz, J., and Kahn, A. (2000) *Org. Electron.*, **1**, 5.

223 Löglund, M., Dannetun, P., Fredricksson, C., Salaneck, W.R., and Bredas, J.L. (1996) *Phys. Rev. B*, **53**, 16327.

224 Hirose, Y., Kahn, A., Aristov, V., Soukiassian, P., Bulovic, V., and Forrest, S.R. (1996) *Phys. Rev. B*, **54**, 13748.

225 Azuma, Y., Akatsuka, S., Okudaira, K.K., Harada, Y., and Ueno, N. (2000) *J. Appl. Phys.*, **87** (2), 766.

226 Wüsten, J., Berger, S., Heimer, K., Lach, S., and Ziegler, Ch. (2005) *J. Appl. Phys.* **98**, 013705

227 Vázquez, H., Oszwaldowski, R., Pou, P., Ortega, J., Pérez, R., Flores, F., and Kahn, A. (2004) *Europhys. Lett.* **65**, 802

228 Ishii, H., Sugiyama, K., Yoshimura, D., Ito, E., Ouchi, Y., and Seki, K. (1998) *IEEE J. Sel. Top. Quantum Electron.*, **4**, 24.

229 Pfeiffer, M., Beyer, A., Fritz, T., and Leo, K. (1998) *Appl. Phys. Lett.*, **73**, 3202.

230 Chasse, T., Wu, C.I., Hill, I.G., and Kahn, A. (1999) *J. Appl. Phys.*, **85**, 6589.

231 Gorelsky, S. I., *AOMix: Program for Molecular Orbital Analysis*, http://www.sg-chem.net/, University of Ottawa, version 6.4, 2010.

232 Zahn, D.R.T., Gavrila, G.N., and Salvan, G. (2007) *Chem. Rev.*, **107**, 1161.

233 Kera, S., Setoyama, H., Onoue, M., Okudaira, K.K., Harada, Y., and Ueno, N. (2001) *Phys. Rev. B*, **63**, 15204.

234 Hirose, K., Foxman, E., Noguchi, T., and Uda, M. (1990) *Phys. Rev. B*, **41**.

235 Evans, D.A., Steiner, H.J., Vearey-Roberts, A.R., Dhanak, V., Cabailh, G., O'Brien, S., McGovern, I.T., Braun, W., Kampen, T.U., Park, S., and Zahn,

D.R.T. (2003) *Appl. Surf. Sci.*, **212–213**, 417–422.

236 Hill, I.G., and Kahn, A. (1998) *Proc. SPIE*, **3476**, 168.

237 Hirose, Y., Kahn, A., Aristov, V., and Soukiassian, P. (1996) *Appl. Phys. Lett.*, **68**, 217.

238 Hirose, Y., Wu, C.I., Aristov, V., Soukiassian, P., and Kahn, A. (1997) *Appl. Surf. Sci.*, **113/114**, 291.

239 Umbach, E., and Fink, R. (2002) Organic nanocrystals: science and applications, in *Proceedings of the International School of Physics "Enrico Fermi", Course CXLIX* (eds V.M. Agranovich and G.C. La Rocca), IOS Press, Netherlands, p. 233.

240 Dweydari, A.W., and Mee, C.H.B. (1973) *Phys. Status Solidi A*, **17**, 247.

241 Dweydari, A.W., and Mee, C.H.B. (1971) *Phys. Status Solidi A*, **27**, 223.

242 Narioka, S., Ishii, H., Edamatsu, K., Kamiya, K., Hasegawa, S., Ueno, N., and Seki, K. (1995) *Phys. Rev. B*, **52**, 2362.

243 Miyamae, T., Hasegawa, S., Yoshimura, D., Ishii, H., Ueno, N., and Seki, K. (2000) *J. Chem. Phys.*, **112**, 333.

244 Ueno, N., Seki, K., Sato, N., Fujimoto, H., Kuramochi, T., Sugita, K., and Inokuchi, H. (1990) *Phys. Rev. B*, **41**, 1176.

245 Gensterblum, G., Pireaux, J.J., Thiry, P.A., Caudano, R., Buslaps, T., Johnson, R.L., Le Lay, G., Aristov, V., Günther, R., Taleb Ibrahimi, A., Indlekofer, G., and Petroff, Y. (1993) *Phys. Rev. B*, **48**, 14756.

246 Yamane, H., Kera, S., Okudaira, K.K., Yoshimura, D., Seki, K., and Ueno, N. (2003) *Phys. Rev. B*, **68**, 033102.

247 Gavrila, G.N., Mendez, H., Kampen, T.U., Zahn, D.R.T., Vyalikh, D.V., and Braun, W. (2004) *Appl. Phys. Lett.*, **85**, 4657.

248 Pireuax, J.J., Riga, J., Caudano, R., Verbist, J.J., Andre, J.M., Delhalle, J., and Delhalle, S., (1974) *J. Electron Spectrosc. Relat. Phenom.*, **5**, 531.

249 Komolov, S.A. (1992) *Total Current Spectroscopy of Surfaces*, Gordon and Breach, Philadelphia, PA.

250 Pendry, J.B. (1974) *Low Energy Electron Diffraction*, Academic Press, London.

251 Komolov, S.A., Gerasimova, N.B., Morozov, A.O. (1996) *Phys. Low-Dim. Struct.*, **11/12**, 81.

252 Strocov, V.N., Starnberg, H.I., Nilsson, P.O., Brauer, H.E., Holleboom, L.J. (1998) *J. Phys.: Condens. Matter* **10**, 5749.

253 [20] Sato, N., Yoshida, H., Tsutsumi, K. (1998) *J. Electron Spectrosc.Relat. Phenom.*, **88–91**, 861.

254 Taborski, J., Vaterlein, P., Dietz, H., Zimmermann, U., Umbach, E. (1995) *J. Electron Spectrosc. Relat. Phenom.*, **75**, 129.

255 Morozov, A.O., Kampen, T.U., and Zahn, D.R.T. (2000) Surface *Science*, **446**, 193.

256 Dai, D., Wang, X., Hu, J., Ge, Y. (1992) *Surf. Sci.*, **274**, 252.

257 Strocov, V.N., Mankefors, S., Kanski, P.O., Ilver, L., Starnberg, H.I. (1999) *Phys. Rev. B*, **59**, R5296.

258 Komolov, S.A. (1981) *Soviet. Phys. Tech. Phys.*, **26**, 1108.

259 Schafer, I., Schluter, M., Skibowski, M. (1987) *Phys. Rev. B*, **35**, 7663.

260 Wright, J.S. (1995) *Molecular Crystals*, 2nd edn, Cambridge University Press.

261 Dimitrakopoulos, C.D., and Mascaro, D.J. (2001) *IBM J. Res. Dev.*, **45**, 11.

262 Pope, M., and Swenberg, C.E. (1999) *Electronic Processes in Organic Crystals and Polymers*, 2nd edn, Oxford Science Publications, Oxford, UK.

263 Karl, N., Kraft, K.-H., Marktanner, J., Münch, M., Schatz, F., Stehle, R., and Uhde, H.-M. (1999) *J. Vac. Sci. Technol. A*, **17**, 2318.

264 Kampen, T.U., Park, S., and Zahn, D.R.T. (2002) *Appl. Surf. Sci.*, **190**, 461.

265 Rhoderick, E.H., and Williams, R.H. (1988) *in Metal-Semiconductor Contacts*, 2nd edn, Clarendon, Oxford.

266 Park, S., Kampen, T.U., Zahn, D.R.T., and Braun, W. (2001) *Appl. Phys. Lett.*, **79**, 4124.

267 Hudej, R., Zavrtanik, M., Brownwell, J.N., and Bratina, G. (2001) *Mater. Technol.*, **35**, 151.

268 Marktanner, J. (1995) Ladungsträgerbeweglichkeiten in dünnen organischen Photoleiter- und Halbleiter-Aufdampfschichten, PhD thesis, Stuttgart.

269 Bolognesi, A., Carlo, A.Di, Lugli, P., Kampen, T.U., and Zahn, D.R.T. (2003) *J. Phys., Condens. Matter*, **15**, S2719.

270 Serena, P.A., Soler, J.M., and Garcia, N. (1986) *Phys. Rev. B*, **34**, 6767.

271 Bolognesi, A., Carlo, A.D.I., and Lugli, P. (2002) *Appl. Phys. Lett.*, **81**, 4646.

272 Bolognesi, A., Carlo, A.Di, Lugli, P., and Conte, G. (2003) *Synth. Met.*, **138**, 95.

273 Bässler, H. (1993) *Phys. Stat. Sol. B*, **175**.

274 Hartenstein, B., Bässler, H., Heun, S., Borsenberger, P., Van der Auweraer, M., and De Schryver, F.C. (1995) *Chem. Phys.*, **191**, 321.

275 Thurzo, I., Kampen, T.U., and Zahn, D.R.T. (2002) *J. Vac. Sci. Technol. A*, **20**, 1597.

276 Szuber, J., and Grzadziel, L. (2000) *Thin Solid Films*, **376**, 214.

277 Szuber, J., and Grzadziel, L. (2001) *Thin Solid Films*, **391**, 282.

278 Kim, H.H., Miller, T.M., Westerwick, E.H., Kim, Y.O., Kwock, E.W., Morris, M.D., and Cerullo, M. (1994) *J. Lightwave Technol.*, **12**, 2107.

279 Li, F., Tang, H., Anderegg, J., and Shinar, J. (1997) *Appl. Phys. Lett.*, **70**, 1233.

280 Kampen, T.U., Bekkali, A., Thurzo, I., Zahn, D.R.T., Bolognesi, A., Ziller, T., Di Carlo, A., and Lugli, P. (2004) *Appl. Sur. Sci.*, **243** (1–4), 313–320.

281 Bekkali, A., Thurzo, I., Kampen, T.U., and Zahn, D.R.T. (2004) *Appl. Sur. Sci.*, **234**, (1–4), 149–154.

282 Yang, J., and Shen, J. (1999) *J. Appl. Phys.*, **85**, 2699.

283 Yakimov, A.V., Savvate'ev, V.N., and Davidov, D. (2000) *Synth. Met.*, **115**, 51.

284 Yamada, T., Zou, D., Jeong, H., Akaki, Y., and Tsutsui, T. (2000) *Synth. Met.*, **111–112**, 237.

285 Thurzo, I., and Gmucová, K., (1994) *Rev. Sci. Instrum.*, **65**, 2244.

286 Mego, T. J., (1986) *Rev. Sci. Instrum.*, **58**, 2798.

287 Thurzo, Ilja, Kampen, Thorsten U., and Zahn, Dietrich R. T., (2002) *J. Vac. Sci. Technol. A*, **20**, 1597.

288 Scaife, B. K. P., (1998) *Principles of Dielectrics*, Clarendon, Oxford.

289 Pons, D., (1984) *J. Appl. Phys.*, **55**, 3644.

290 Herbst, W. Hūnger, K.: *Indūstrielle Organische Pigmente*, VCH 1995, 2nd edition.

291 St. Hohenecker, Chalcogen modification of GaAs(100) surface and metal/GaAs(100) contacts, PhD thesis, Chemnitz University of Technology, 2001.

292 Friedrich, M., Gavrila, G., Himcinschi, C., Kampen, T.U., Kobitski, A.Yu, Méndez, H., Salvan, G., Cerrilló, I., Méndez, J., Nicoara, N., Baró, A.M., and Zahn, D.R.T. (2003) *J. Phys. Condens. Matt.*, **15**, S2699.

294 Meyer zu Heringdorf, F.-J., Reuter, M.C., Tromp, R.M. (2001) "Growth Dynamics of Pentacene Thin Films", *NETURE*, **412**, 517–520.

295 Meyer zu Heringdorf, F.-J., Reuter, M.C., Tromp, R.M. (2004) "The nucleation of pentacene thin films", *Applied Physics A*, **78**, 787–791.

Index

Low Molecular Weight Organic Semiconductors. Thorsten U. Kampen

ISBN: 978-3-527-40653-1